Nelson Advanced Science

Make the Grade Chemistry

AS and A2

George Facer

Published in 2000 by:
Nelson Thornes Ltd

Second edition published in 2003 by:
Nelson Thornes Ltd
Delta Place
27 Bath Road
CHELTENHAM
GL53 7TH
United Kingdom

03 04 05 06 / 10 9 8 7 6 5 4 3 2 1

A catalogue record for this book is available from the British Library

ISBN 0 7487 7281 2

Illustrations and page make-up by Hardlines and Mathemetical Composition Setters Ltd.

Printed in Croatia by Zrinski

The practice questions and mark schemes are based upon existing Edexcel Foundation questions.

Cover photograph credit: Photodisc 72 (NT)

Every effort has been made to trace all of the copyright holders, but where this has not been possible the publisher would be pleased to make any necessary arrangements at the first opportunity.

Contents

Introduction

How to use this book

This book has been written to be used alongside the Nelson Advanced Science Chemistry student books. It aims to help you develop your study skills, to make your learning more effective and to give you help with your revision. You may be given Unit Tests at different stages of your course, so you should be prepared right from the beginning.

The Units of this book are arranged in the order of the Units of the specification (or syllabus, as it used to be called). Units 1 to 3 of the book deal with the AS content. Units 4 to 6 will help you to prepare for the A2 assessments if you are staying on to study for Advanced GCE Chemistry.

The Edexcel specification (September 2002) divides AS and A2 Chemistry into a total of twenty two topics, and this revision book deals with each topic separately and in the order in which they occur in the specification as shown in the Contents list.

The format in each topic is:

- **An introduction** – gives an overview of important concepts which provide a key to the understanding of the topic being covered

- **Things to learn** – these are mainly definitions

- **Things to understand** – includes worked examples and diagrams

- **A checklist** – to check that you have covered and have understood all that is in the topic

- **Testing your knowledge and understanding** – this consists of questions that either require short answers, which are given in the margin, or those that require longer answers, which appear in the back of the book.

There are **practice test papers** at the end of each chapter. The format and length are similar to those in the actual Unit Tests. The questions in the practice tests for chapters 1, 2, 4, 5 and 6 are taken from previous Edexcel AS or A2 papers.

The Edexcel specification will be tested in three units for AS and three for A2. However Unit Test 3 is split into two parts – 3A testing practical skills and 3B which is a written paper that tests the theory underpinning practical chemistry.

> Unit Test 3A will either be assessed by your school or college using the first year's practical work or will consist of a $1^3/_4$ hour practical test set and marked by Edexcel.

Unit Test	Type	Duration	AS or A2
1	Structured questions	1 hr	AS
2	Structured questions	1 hr	AS
3A	Coursework assessment or practical test	1st year's work 1 hr 45 min	AS
3B	Structured questions	1 hr	AS
4	Structured questions	1 hr 30 min	A2
5	Structured questions with some synoptic issues	1 hr 30 min	A2
6A	Coursework assessment or practical test	2nd year's work 1 hr 45 min	A2
6B	Section A: analysis of practical data Section B: synoptic	1 hr 30 min	A2

A **synoptic** question is one that requires the drawing together of knowledge, understanding and skills learned in different parts of the Advanced GCE course. The emphasis will be on understanding and applying principles. The questions will require you to make connections between different areas of chemistry and cope with compounds that are new to you. You will often be expected to use the data in the question to deduce what happens rather than to recall it.

Do not leave your revision until the last minute. Revision should take place throughout the whole course.

Make the Grade in AS and A2 Chemistry summarises the techniques that should be mastered for the practical and includes a selection of questions similar to those that will be set for Unit Test 3B.

Similarly Unit Test 6 is split into two parts. As before, Unit Test 6A (either assessed by your school/college on your second year's work or taken as a practical exam lasting 1 hour 45 minutes) tests practical skills, especially qualitative ones. Unit Test 6B consists of two sections and lasts for 1 hour 30 minutes. Section A will test analysis of experimental data and section B consists of three **synoptic** questions. This book will help you to tackle this paper and includes a sample test paper.

A mark scheme, together with a table of the percentages needed for each grade, is provided in the Answers section (pages 121–142). A Periodic Table is provided on the inside back cover of the book.

Study skills

Revision is a personal activity. What works best for you may not be so effective for someone else. However there are some golden rules:

1 Revise little and often.

2 Revise actively – do not sit and stare at your notes or this book. Write down important points or use a highlighter to mark important passages in your notes or in this book (but only if you own it!).

3 Write down answers to the topic questions and then check them with those given.

4 Help each other. Explaining a point of chemistry to another student is a good way of clarifying your own understanding. Test each other by asking simple questions, such as formulae, definitions and reagents and products of organic reactions.

Here are some suggestions to help you study and prepare for your Unit Test papers:

Daily tasks

After each lesson check that your notes are complete. Try spending 10 to 15 minutes looking through them. If there is something that you do not understand:

● Read the relevant part in this book or your text book and, if necessary, add to your notes so that they will be clear when you read them again.

● Discuss the problem with another student.

● If you still have difficulty, ask your teacher as soon as you can.

● If you can solve the problem by yourself, your understanding will often be deeper and longer lasting.

Weekly tasks

● Go through your notes. Highlight important parts.

● Read through the relevant parts of this book and make notes and/or highlight important points.

● Complete any homework assignments.

End of topic tasks

When your teacher has completed a topic, you should revise that topic thoroughly. To do this:

● Work through your notes with a copy of the Edexcel specification (syllabus).

Helpful hints

The specification can be found on Edexcel's web site www.edexcel.org.uk

- Summarise your notes to the bare essentials.
- Work through the topic in this book. Discuss any difficulties with other students.
- Write answers to **all** the questions in the 'testing your knowledge and understanding' section of the topic.

Preparing for the Unit Tests (examinations)

If you have followed the advice above, you will be well prepared for the assessment tests. If you are taking the three AS units they will all be on the same day, so you will have to start your revision in plenty of time. Units 4 and 5 will be in a single session and 6B will be about a week later.

- Try spending about 30 minutes revising one subject. Then switch from chemistry to another subject.
- Take regular breaks.
- Revise actively with pen, highlighter and paper.

When you have revised all the topics in a unit, sit down for 1 hour at AS and 1 hour and 30 minutes at A2 and do the relevant practice Unit Test provided. These are found at the end of each chapter. Mark the test – or better still mark a friend's answers and let him or her mark yours. Then:

- Work out where you went wrong.
- If you obtained low marks for a particular topic, go back to your notes and to this book and rework that topic.

Spreading revision in this way over the whole course will reduce stress and will guarantee a better grade than you would obtain by leaving it all to a mad dash at the end. Chemistry is a subject in which knowledge is built up gradually. This is particularly true of organic chemistry, which many students find to be a difficult topic until they suddenly get a feel for it and it all falls into place. The more thoroughly you work in the earlier stages, the easier and more enjoyable you will find the study of chemistry.

The day of the Unit Test

If you have followed the advice given here, you should feel confident that you will be able to do your best. Some people find it helpful to spend a little time looking over some chemistry before going into the test. Check that you have:

- Two or more blue or black pens and several pencils
- Your calculator – if the batteries are old replace them beforehand
- A watch – try putting it on the desk in front of you
- A ruler
- A good luck charm, if it helps.

Don't take a red pen with you as the awarding body doesn't allow you to use this colour – the examiners use red for marking the papers.

Tackling the question paper

- Work steadily through the paper starting at question 1.
- All questions (except Unit Test 6B) are of the structured type, where you answer the question on the question paper itself.
- If you need more room for your answer, look for space at the bottom of the page, the end of the question or after the last question.

Helpful hints

If you use space at the bottom of the page let the examiner know by adding 'continued below' or 'continued on page XX'.

- Use the amount of space given for each answer as a guide to how much you should write. If a question has three lines for the answer do not write an essay. Work out the essential points that should be made and check them against the number of marks to be awarded.
- Do not repeat the question in your answer.
- Pace yourself so that you neither run out of time nor have masses to spare at the end. If you get stuck, do not waste time. Make a note of the question number and part which caused you difficulty and go on. Later, if you have time, go back and try that part again.
- Do **not** use correcting fluid. It is forbidden by the Board's regulations. Instead neatly cross out what you have written. If, later, you realise that what you first wrote is correct, write 'ignore crossing out' by the work that you had crossed out. The examiner will then mark it.

Unit Test 5

This unit test will examine the content of Unit 5 and will consist of a number of structured questions. You must answer all questions, some of which will be synoptic. This will require you to use skills and knowledge from earlier units and this applies especially to organic questions, which will require knowledge of the material covered in topics 2.2 and 4.5.

Unit Test 6B

All the questions in this paper are synoptic and are based on the entire AS and A2 specification. Section A is compulsory and will test your ability to analyse experimental data – often of a reaction new to you. Section B will consist of three questions and you have to answer two of them. Spend some time looking through these questions before deciding which two you will attempt. If you find that you are getting nowhere in a question, abandon it and try another, but **do not cross out what you have written** – all your work will be marked and you will be awarded the marks for your two highest scoring questions. Do not waste time worrying about the fact that a situation is unfamiliar to you. This is the point of synoptic questions. The questions themselves give you all the clues that you need to relate the new material to work that you have covered in your A-level course. Remember that the questions have a thread and lead you from one stage to the next.

Terms used in the Tests

You should understand what the examiners want. Terms that are often used in the question papers are explained below:
- **Define**: It is often helpful to include an example or equation to supplement your definition.
- **Formula**: A molecular formula is sufficient.
- **Structural formula**: This must be unambiguous. It is acceptable to use CH_3 and C_2H_5 but not C_3H_7. If in doubt write a full structural formula.
- **Full structural formula**: All atoms and all bonds must be shown.
- **State**: No explanation is required, nor should one be given.
- **Explain (or why is?)**: Look to see how many marks there are for the question. Make sure that you give at least the same number of pieces of information in your answer.
- **Deduce**: Use the data supplied to answer the question.
- **Hence deduce**: Use the answer that you have obtained in the previous section to work out the answer.

Helpful hints

As a rough guide a question out of 15 marks should take about 15 minutes – a rate of 1 minute per mark.

Helpful hints

Formulae with methyl groups written as CH_3 will usually be acceptable, especially in questions on stereoisomerism. Don't forget to draw in all the hydrogen atoms and check that each carbon has four bonds, each oxygen two and the hydrogens only one bond.

- **Suggest**: You are not expected to have learnt the answer. You should be able to apply your understanding of similar substances or situations in order to work out the answer.
- **Compare**: You should make a valid comment about **both** substances.
- **Calculate**: It is essential to show your working and to set your work out clearly. If you do, you can score many of the marks even if your final answer is wrong because you made a slip, part of the way through the calculation. Give your answer to the same number of significant figures as there are in the data.
- **Reagents**: Full names (as on the bottle of the substance) or full formulae are required.
- **Conditions**: If room temperature is the correct condition, it should be stated. The term **reflux** is not taken to imply heating, so the term **heating under reflux** should be used.
- **Equations**: These should be balanced. Ionic equations should be used where appropriate. In organic reactions, the use of [H] and [O] to represent reducing and oxidising agents is acceptable.

  Do not give a mechanism for organic reactions if an overall equation is asked for.

  State symbols should be given:
  - in all thermochemical equations
  - in all electrochemical equations
  - where a precipitate or gas is produced
  - whenever they are asked for in the question.
- **Outline the preparation of**: Only in this type of question may a flow diagram be used, with names of reagents and conditions written on the arrow.
- **Stable**: This word should always be qualified, e.g. 'stable to heat', or 'the reactants are thermodynamically stable compared with the products'.

and finally

Examiners **do** try, wherever possible, to give you marks rather than looking for ways to take them away.

Be prepared, be confident and you will do your best, which is all that anyone can ask of you.

Resource materials from Edexcel

- *Formulae, Moles and Equations – A Workbook*
- *The Edexcel AS and Advanced GCE in Chemistry Externally Assessed Practical – A User Guide.*

Both of these can be obtained from Edexcel Publications, Adamsway, Mansfield, Notts NG18 4LN or e-mail: publications@maillin.co.uk

Acknowledgements

I would like to thank the staff at Edexcel, and especially Ray Vincent, for their help and advice, and Rod Beavon, Brian Chapman, Angela Melamed and Philip Eastwood for their help in the transformation of this book from first draft to its final form. I am grateful to my wife, Judy, for her patience, and to my younger son, Michael, for his great help in showing me how to get the best from my computer and the Net.

The author George Facer is former Head of Science at Sherborne School and is a Principal Examiner in Chemistry for the Edexcel Foundation.

1 Structure, bonding and main group chemistry

Topic 1.1   Atomic structure

Introduction

Much of chemistry depends upon Coulomb's Law which states that the electrostatic force of attraction, F, is given by:

$$F \propto (q_+.q_-) \div r^2$$

where q_+ and q_- are the charges on the objects (e.g. the nucleus, an electron, ions etc.) and r^2 is the square of the distance between their **centres**. This means that the bigger the charge, the bigger the force, and the further the centres are apart, the weaker the force.

Things to learn

- **Atomic number (Z)** of an element is the number of protons in the nucleus of its atom.

- **Mass number** of an isotope is the number of protons plus the number of neutrons in the nucleus.

- **Isotopes** are atoms of the same element which have the same number of protons but different numbers of neutrons. They have the same atomic number but different mass numbers.

- **Relative atomic mass (A_r)** of an element is the **average** mass (taking into account the abundance of each isotope) of the atoms of that element relative to 1/12th the mass of a carbon-12 atom.

- **Relative isotopic mass** is the mass of one atom of an isotope relative to 1/12th the mass of a carbon-12 atom.

- **Relative molecular mass (M_r)** of a substance is the sum of all the relative atomic masses of its constituent atoms.

- **Molar mass** is the mass of one mole of the substance. Its units are grams per mole ($g\,mol^{-1}$), and it is numerically equal to the relative molecular mass.

- **1st ionisation energy** is the amount of energy required per mole to **remove** one electron from each gaseous atom to form a singly positive ion

$$E(g) \rightarrow E^+(g) + e^-$$

- **2nd ionisation energy** is the energy change per mole for the removal of an electron from a singly positive gaseous ion to form a doubly positive ion

$$E^+(g) \rightarrow E^{2+}(g) + e^-$$

The relative molecular mass is also called the relative formula mass especially for ionic substances.

Helpful hints

Ionisation energies are always endothermic and relate to the formation of a positive ion.

☐ **1st electron affinity** is the energy change per mole for the **addition** of one electron to a gaseous atom to form a singly negative ion

$$E(g) + e^- \rightarrow E^-(g)$$

☐ **2nd electron affinity** is the energy change per mole for the addition of an electron to a singly negative gaseous ion to form a doubly negative ion

$$E^-(g) + e^- \rightarrow E^{2-}(g)$$

☐ **s block** elements are those in which the highest occupied energy level is an s orbital. They are in Groups 1 and 2.

Similar definitions apply to **p block** (Groups 3 to 7 and 0) and **d block** (Sc to Zn) elements.

 Things to understand

Mass spectra

● An element is first vapourised and then bombarded by high-energy electrons that remove an electron from the element and form a positive ion. This ion is then accelerated through an electric potential, deflected according to its mass and finally detected.

● Metals and the noble gases form singly positively charged ions in the ratio of the abundance of their isotopes.

● Non-metals also give molecular ions. For example Br_2, which has two isotopes ^{79}Br (50%) and ^{81}Br (50%), will give three lines at m/e values of 158, 160 and 162 in the ratio 1:2:1. These are caused by $(^{79}Br–^{79}Br)^+$, $(^{79}Br–^{81}Br)^+$ and $(^{81}Br–^{81}Br)^+$.

● The relative atomic mass of an element can be calculated from mass spectra data as follows:

A_r = the sum of (mass of each isotope × percentage of that isotope)/100

Helpful hints

A common error is to miss out the + charge on the formula of a species responsible for a line in a mass spectrum.

Worked example

Boron was analysed in a mass spectrometer.
Calculate the relative atomic mass of boron using the results below.

Peaks at m/e of	Abundance (%)
10.0	18.7
11.0	81.3

Answer: A_r = (10.0 x 18.7 + 11.0 x 81.3) / 100 = 10.8

Electron structure

● The first shell only has an s orbital.
● The second shell has one s and three p orbitals.
● The third and subsequent shells have one s, three p and five d orbitals.
● Each orbital can hold a maximum of **two** electrons.
● The order of filling orbitals is shown in Figure 1.1 below.

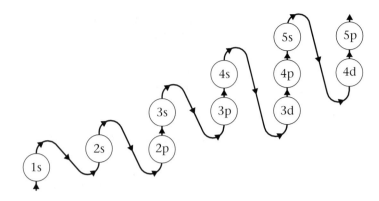

Fig 1.1 *The order of filling of atomic orbitals*

Electron structures can be shown in two ways:

● The s, p, d notation. For vanadium (atomic number 23) this is:
   $1s^2, 2s^2\ 2p^6, 3s^2\ 3p^6\ 3d^3, 4s^2$.

● The electrons in a box notation. For phosphorus ($Z = 15$) this would be:

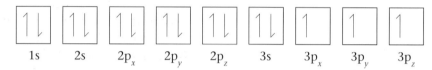

| 1s | 2s | $2p_x$ | $2p_y$ | $2p_z$ | 3s | $3p_x$ | $3p_y$ | $3p_z$ |

Sizes of atoms and ions

● The atoms become **smaller** going across a period from left to right, because the nuclear charge increases, pulling the electrons in closer, though the number of shells is the same.

● The atoms get **bigger** going down a group, because there are more shells of electrons.

● A positive ion is smaller than the neutral atom from which it was made, because the ion has one shell fewer than the atom.

● A negative ion is bigger than the neutral atom, because the extra repulsion between the electrons causes them to spread out.

1st ionisation energy

● There is a general increase going from left to right across a period (see Figure 1.2). This is caused mainly by the increased nuclear charge (atomic number) without an increase in the number of inner shielding electrons.

> ### Helpful hints
>
> In an atom, the outer electrons are shielded from the pull of the nucleus by the electrons in shells nearer to the nucleus (the inner electrons).

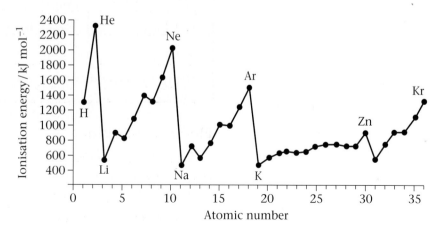

Fig 1.2 *The first ionisation energies (kJ mol^{-1}) of the elements up to krypton*

You should be able to sketch the variation of 1st ionisation energies with atomic number for the first 20 elements.

The first six ionisation energies/ kJ mol^{-1} of an element are:

1st   786
2nd   1580
3rd   3230
4th   4360
5th   16000
6th   20000

There is a big jump after the fourth, so the element is in Group 4.

- There are slight decreases after Group 2 (this is because, for Group 3, it is easier to remove an electron from the higher energy p orbital), and after Group 5 (this is because, for Group 6, the repulsion of the two electrons in the p_x orbital makes it easier to remove one of them).
- There is a decrease going down a Group. This is caused by the outer electron being further from the nucleus. (The extra nuclear charge is balanced by the same extra number of inner shielding electrons.)

Successive ionisation energies

- The 2nd ionisation energy of an element is always bigger than the first, because the second electron is removed from a positive ion.
- When there is a very big jump in the value of successive ionisation energies, an electron is being removed from a lower shell, e.g. if this jump happens from the 4th to the 5th ionisation energy, four electrons have been removed from the outer shell during the first four ionisations, and so the element is in Group 4.

Electron affinity

- The 1st electron affinity values are always negative (exothermic), as a negative electron is being brought towards the positive nucleus in a neutral atom. They are the most exothermic for the halogens.
- The 2nd electron affinity values are always positive (endothermic), because a negative electron is being added to a negative ion.

 Checklist

Before attempting the questions on this topic, check that you can:

❑ Define A_r, M_r, and relative isotopic mass.

❑ Calculate the number of neutrons in an isotope, given the atomic and mass numbers.

❑ Calculate the relative atomic mass of an element from mass spectra data.

❑ Define first and subsequent ionisation energies.

❑ Explain the changes in first ionisation energies for elements across a Period and down a Group.

❑ Deduce an element's Group from successive ionisation energies.

❑ Work out the electronic structure of the first 36 elements.

❑ Define the 1st and 2nd electron affinities.

Testing your knowledge and understanding

Answers
Mass 1, charge +1
Mass 1, charge 0
Mass 1/1860 , charge –1
23 (p+n) – 11p = 12n
99 (p+n) – 56n = 43.
$Na(g) \rightarrow Na^+(g) + e^-$
$Cl(g) \rightarrow Cl^+(g) + e^-$
$Mg^+(g) \rightarrow Mg^{2+}(g) + e^-$
$1s^2, 2s^2, 2p^2$
$1s^2, 2s^2, 2p^6, 3s^2.$
$1s^2, 2s^2, 2p^6, 3s^2, 3p^6, 3d^6, 4s^2$

 The answers to the numbered questions are on page 121.

For the following set of questions, cover the margin, write down your answer, then check to see if you are correct. (You may refer to the Periodic Table on the inside back cover)

● State the masses and charges (relative to a proton) of:
  a proton (p)
  a neutron (n)
  an electron (e).

● How many neutrons are there in an atom of $^{23}_{11}Na$?

● What is the atomic number of an element which has an atom of mass number 99 and contains 56 neutrons ?

● Write equations with state symbols for:
  a the 1st ionisation energy of sodium
  b the 1st ionisation energy of chlorine
  c the 2nd ionisation energy of magnesium.

● Using the 1s, 2s 2p . . notation, give the electronic structures of the elements:
  a $_6C$
  b $_{12}Mg$
  c $_{26}Fe$.

1 Explain the difference between relative isotopic mass and relative atomic mass. Illustrate your answer with reference to a specific element.

2 Lithium has naturally occurring isotopes of mass numbers 6 and 7.
   Explain why its relative atomic mass is 6.9 not 6.5.

3 Magnesium was analysed in a mass spectrometer.
   Peaks were found at three different *m/e* ratios.

m/e	Abundance (%)
24.0	78.6
25.0	10.1
26.0	11.3

   Calculate the relative atomic mass of magnesium.

4 Gallium has a relative atomic mass of 69.8. Its mass spectrum shows two peaks at *m/e* of 69.0 and 71.0.
   Calculate the percentage of each isotope in gallium.

5 The successive ionisation energies of an element X are given below:

Ionisation energy	1st	2nd	3rd	4th	5th	6th	7th	8th
Value/kJ mol^{-1}	1060	1900	2920	4960	6280	21 200	25 900	30 500

   State, giving your reasons, which group the element X is in.

6 **a** Sketch the variation of 1st ionisation energy against atomic number for the first 11 elements.

  b Explain why the ionisation energy is less for:
    i the element of atomic number (*Z*) of 5 than that for element of
       Z = 4
    ii the element of *Z* = 8 than that for element of *Z* = 7

iii the element of $Z = 11$ than that for element of $Z = 3$.

7 Using the electron in a box notation, give the electronic structure of chlorine ($Z = 17$).

8 Explain, in terms of electronic structure, why the chemical properties of lithium, sodium and potassium are similar but not identical.

9 a Give the equation, with state symbols, that represents the 1st electron affinity of **i** lithium, **ii** chlorine, **iii** oxygen.

b Give the equation that represents the 2nd electron affinity of oxygen.

c Why is the 2nd electron affinity of oxygen endothermic whereas the 1st electron affinities of lithium, chlorine and oxygen are all exothermic?

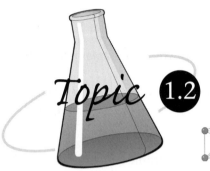

Topic **1.2** Formulae, equations and moles

 ## *Introduction*

The keys to this topic are:
- To be able to calculate the number of moles from data.
- To set out calculations clearly.

 ## *Things to learn*

☐ The Avogadro Constant is the number of carbon atoms in exactly 12 g of the carbon-12 isotope. Its value is 6.02×10^{23} mol^{-1}.

☐ One mole of a substance is the amount of that substance that contains 6.02×10^{23} particles of that substance. This means that one mole of a substance is its relative atomic or molecular mass expressed in grams.

> 1 mol of NaOH has a mass of 40 g
> 1 mol of O_2 has a mass of 32 g.

☐ The molar mass of a substance is the mass (in grams) of one mole.

☐ **Amount** of substance is the number of moles of that substance.

☐ The empirical formula is the simplest whole number ratio of the elements in the compound.

 ## *Things to understand*

Calculation of empirical formulae from percentage data

It is best calculated using a table.

Element	Percentage of element	Percentage divided by A_r of element	Divide each by lowest
Carbon	48.7	48.7 ÷ 12 = 4.1	4.1 ÷ 2.7 = 1.5
Hydrogen	8.1	8.1 ÷ 1 = 8.1	8.1 ÷ 2.7 = 3
Oxygen	43.2	43.2 ÷ 16 = 2.7	2.7 ÷ 2.7 = 1

The last column gives the empirical formula, but if any value in this column comes to a number ending in .5 or .25, you must multiply all the values by 2 or 4 to get integers. So here the empirical formula is $C_3H_6O_2$.

Equations

These must balance. The number of atoms of an element on one side of the equation must be the same as the number of atoms of that element on the other side.

Ionic equations

There are three rules:

- Write the ions separately for solutions of ionic compounds (salts, strong acids and bases).
- Write full 'molecular' formulae for solids and all covalent substances.
- Spectator ions must be cancelled and so do not appear in the final equation.

Helpful hints

Ionic equations must also balance for charge.

Moles

There are three ways of calculating the amount of substance (in moles):

- For a pure substance X

   The amount of X (in moles) = mass of X (in grams)/its molar mass

Helpful hints

Avoid writing mol = 0.11. Instead state the name or formula of the substance
 e.g. amount of Na = 0.11 mol.

Worked example

Calculate the amount of H_2O in 1.1 g of water

Answer: 1.1 g/18 g mol^{-1} = 0.061 mol of H_2O

- For solutions:

   The amount of solute = concentration (in mol dm^{-3}) x volume (in dm^3)

   Therefore concentration = moles / volume in dm^3

Helpful hints

As volume in dm^3 = volume in cm^3/1000, the following formula can be used:
moles = M x V/1000
or concentration = moles x 1000/V
where M is the concentration and V is the volume in cm^3.

Worked example

Calculate the amount of NaOH in 22.2 cm^3 of 0.100 mol dm^{-3} solution

Answer: 0.100 x 0.0222 = 2.22 x 10^{-3} mol of NaOH

- For gases:

   Amount of gas (in moles) = volume (in dm^3)/molar volume.

   The **molar volume** of a gas is 24 dm^3 mol^{-1}, measured at room temperature and pressure (RTP).

> ## Worked example
>
> Calculate the amount of $H_2(g)$ in 3.2 dm^3 at RTP.
>
> *Answer*: 3.2/24 = 0.13 mol of $H_2(g)$

Calculation of number of particles

The number of particles can be calculated from the number of moles.

● The number of molecules = moles × Avogadro's constant.
● The number of ions = moles × Avogadro's constant × the number of those ions in the formula.

> ## Worked example
>
> Calculate the number of carbon dioxide molecules in 3.3 g of CO_2.
>
> *Answer*: Amount of CO_2 = 3.3/44 = 0.075 mol
> Number of molecules = 0.075 x 6.02 x 10^{23} = 4.5 x 10^{22}
>
> Calculate the number of sodium ions in 5.5 g of Na_2CO_3
>
> *Answer*: Amount of Na_2CO_3 = 5.5/106 = 0.0519 mol
> Number of Na^+ ions = 0.0519 x 6.02 x 10^{23} x 2 = 6.2 x 10^{22}

Calculations based on reactions

These can only be done if a correctly balanced equation is used.

Reacting mass questions

First write a balanced equation for the reaction.
Then follow the route:

```
            Step 1            Step 2            Step 3
Mass A ─────────► Moles A ─────────► Moles B ─────────► Mass B
```

For Steps 1 and 3 use the relationship:

amount of A or B (in moles) = mass/molar mass

For Step 2 use the stoichiometric ratio from the equation:

moles of B = moles of A × ratio B/A

In Step 2, the stoichiometric ratio is 2/1 as there are 2NaOH molecules for each 1SiO_2 molecule in the equation.

> ## Worked example
>
> Calculate the mass of sodium hydroxide required to react with 1.23 g of silicon dioxide.
>
> *Answer*:
> Equation: $SiO_2 + 2NaOH \rightarrow Na_2SiO_3 + H_2O$
> Step 1   amount of SiO_2 = 1.23/60 mol   = 0.0205 mol
> Step 2   amount of NaOH = 0.0205 mol x 2/1 = 0.0410 mol
> Step 3   mass of NaOH = 0.0410 x 40   = 1.64 g

Titrations (this will only be examined at AS in Unit test 3B)

The route is much the same:

	Step 1	Step 2	Step 3	answer
Concentration and volume of A	$\longrightarrow$ Moles A	$\longrightarrow$ Moles B	$\longrightarrow$ about B	

For Steps 1 and 3 use the relationship:

amount (in moles) $= M \times V/1000$

For Step 2 use the stoichiometric ratio from the equation:

moles of B $=$ moles of A $\times$ ratio B/A

Worked example

25.0 cm^3 of a solution of sodium hydroxide of concentration 0.212 mol dm^{-3} was neutralised by 23.4 cm^3 of a solution of sulphuric acid. Calculate the concentration of the sulphuric acid solution.

Answer. Equation: $2NaOH + H_2SO_4 \rightarrow Na_2SO_4 + 2H_2O$

Step 1   amount of NaOH $= 0.212 \times 25.0/1000 = 5.3 \times 10^{-3}$ mol

Step 2   amount of $H_2SO_4 = 5.3 \times 10^{-3} \times 1/2 = 2.65 \times 10^{-3}$ mol

Step 3   concentration of $H_2SO_4 = 2.65 \times 10^{-3}$ mol/0.0234 dm^3

$= 0.113$ mol dm^{-3}

> In Step 2 the stoichiometric ratio is 1/2 as there is 1H_2SO_4 molecule for every 2NaOH molecules in the equation.

Concentration of solutions

This is either: $\dfrac{\text{amount of solute (in moles)}}{\text{volume of solution in dm}^3}$ units: mol dm^{-3}

or: $\dfrac{\text{mass of solute (in grams)}}{\text{volume of solution in dm}^3}$ units: g dm^{-3}

Gas volume calculations

1   For reactions where a gas is produced from solids or solutions, follow:

Step 1	Step 2	Step 3
Mass of A $\longrightarrow$ Moles of A	$\longrightarrow$ Moles of gas B	$\longrightarrow$ Volume of gas B

Step 1 use the relationship:

moles $=$ mass/molar mass

Step 2 use the stoichiometric ratio from the equation:

moles of A $=$ moles of B $\times$ ratio of B/A

Step 3 use the relationship:

volume of gas B $=$ moles of B $\times$ molar volume

Worked example

Calculate the volume of carbon dioxide gas evolved, measured at room temperature and pressure, when 7.8 g of sodium hydrogen carbonate is heated. The molar volume of a gas is 24 dm^3 mol^{-1} at the temperature and pressure of the experiment.

Answer. Equation: $2NaHCO_3 \rightarrow Na_2CO_3 + H_2O + CO_2(g)$
Step 1   amount of $NaHCO_3$ = 7.8/84 mol = 0.0929 mol
Step 2   amount of CO_2 = 0.09286 mol x 1/2 = 0.0464 mol
Step 3   volume of CO_2 = 0.0464 mol x 24 dm^3 mol^{-1} = 1.1 dm^3

> In Step 2 the stoichiometric ratio is 1/2 as there is 1CO_2 molecule for every 2$NaHCO_3$ molecules in the equation.

2   For calculations involving gases only, a short cut can be used. The volumes of the two gases are in the same ratio as their stoichiometry in the equation.

Worked example

What volume of oxygen is needed to burn completely 15.6 cm^3 of ethane?
Answer. Equation: $2C_2H_6(g) + 7O_2(g) \rightarrow 4CO_2(g) + 6H_2O(l)$
Calculation: $\dfrac{\text{volume of oxygen gas}}{\text{volume of ethane gas}} = \dfrac{7}{2} = 3.5$
volume of oxygen gas = 3.5 × 15.6 = 54.6 cm^3

Helpful hints

- You must show all the steps in your calculations
- Don't cut down to two or three significant figures in the middle of a calculation
- Check every calculation to ensure that you have entered the data correctly

Significant figures

You should always express your answer to the same number of significant figures as stated in the question or as there are in the data.

If you cannot work this out in an exam, give your answer to 3 significant figures (or 2 decimal places for pH calculations), and you are unlikely to be penalised. Do not round up numbers in the middle of a calculation. Any intermediate answers should be given to at least 1 more significant figure than your final answer.

 Checklist

Before attempting questions on this topic, check that you can:

☐ Calculate the empirical formula of a substance from the % composition.

☐ Write balanced ionic equations.

☐ Calculate the number of moles of a pure substance from its mass, of a solute from the volume and concentration of its solution, and of a gas from its volume.

☐ Calculate reacting masses and reacting gas volumes.

☐ Use titration data to calculate the volume or the concentration of one solution.

Testing your knowledge and understanding

For the first set of questions, cover the margin, write down your answer, then check to see if you are correct.

● The table below contains data which will help you.

Substance	Solubility
Nitrates	All soluble
Chlorides	All soluble except for AgCl and $PbCl_2$
Sodium compounds	All soluble
Hydroxides	All insoluble except for Group 1 and barium hydroxides

Write ionic equations for the reactions of solutions of:

a lead nitrate and potassium chloride
b magnesium chloride and sodium hydroxide
c sodium chloride and silver nitrate
d sodium hydroxide and hydrochloric acid.

● Calculate the amount (in moles) of:

a Na in 1.23 g of sodium metal
b NaCl in 4.56 g of solid sodium chloride
c Cl_2 in 789 cm^3 of chlorine gas at room temperature and pressure
d NaCl in 32.1 cm^3 of a 0.111 mol dm^{-3} solution of sodium chloride.

● Calculate the volume of a 0.222 mol dm^{-3} solution of sodium hydroxide, which contains 0.0456 mol of NaOH.

● Calculate the number of water molecules in 1.00 g of H_2O.

● Calculate the number of sodium ions in 1.00 g of Na_2CO_3.

● 4.44 g of solid sodium hydroxide was dissolved in water and the solution made up to 250 cm^3. Calculate the concentration in

a g dm^{-3}
b mol dm^{-3}.

Answers

a $Pb^{2+}(aq) + 2Cl^-(aq) \rightarrow PbCl_2(s)$
b $Mg^{2+}(aq) + 2OH^-(aq) \rightarrow Mg(OH)_2(s)$
c $Cl^-(aq) + Ag^+(aq) \rightarrow AgCl(s)$
d $H^+(aq) + OH^-(aq) \rightarrow H_2O(l)$

---

a 1.23/23   = 0.0535 mol
b 4.56/58.5 = 0.0779 mol
c 0.789/24 = 0.0329 mol
d 0.111 x 0.0321 = 3.56 x 10^{-3} mol

0.0456/0.222 = 0.205 dm^3 = 205 cm^3

(1.00/18) x 6.02 x 10^{23} = 3.34 x 10^{22}

(1.00/106) x 6.02 x 10^{23} × 2 = 1.14 x 10^{22}

---

a 4.44 g/0.250 dm^3 = 17.8 g dm^{-3}

b 4.44/40 = 0.111 mol
    Therefore 0.111 mol/0.250 dm^3
    = 0.444 mol dm^{-3}

 The answers to the numbered questions are on pages 121–122.

1 a An organic compound contains 82.76% carbon and 17.24% hydrogen by mass. Calculate its empirical formula.

b It was found to have a relative molecular mass of 58. Calculate its molecular formula.

2 Balance the equations:

a $NH_3 + O_2 \rightarrow NO + H_2O$

b $Fe^{3+}(aq) + Sn^{2+}(aq) \rightarrow Fe^{2+}(aq) + Sn^{4+}(aq)$

3 What mass of sodium hydroxide is needed to react with 2.34 g of phosphoric(V) acid, H_3PO_4, to form the salt Na_3PO_4 and water?

4 What volume of 0.107 mol dm^{-3} potassium hydroxide, KOH, solution is needed to neutralise 12.5 cm^3 of a 0.0747 mol dm^{-3} solution of sulphuric acid?

5 What volume, measured at room temperature and pressure, of hydrogen sulphide gas, H_2S, is required to react with 25 cm^3 of a 0.55 mol dm^{-3}

solution of bismuth nitrate, $Bi(NO_3)_3$? The molar volume of a gas is 24 dm^3 mol^{-1} under these conditions. They react according to the equation:

$$3H_2S(g) + 2Bi(NO_3)_3(aq) \rightarrow Bi_2S_3(s) + 6HNO_3(aq).$$

6 What volume of hydrogen gas is produced by the reaction of 33 dm^3 of methane gas when it is reacted with steam according to the equation:

$$CH_4(g) + H_2O(g) \rightarrow CO(g) + 3H_2(g) ?$$

Structure and bonding

Introduction

You should be able to distinguish between:

● Chemical bonds

Ionic	–	between separate ions
Covalent	–	which are divided into two types:
	1	polar covalent where the bonding pair of electrons is nearer to one atom
	2	pure covalent where the bonding pair of electrons is shared equally
Metallic	–	bonding caused by electrons delocalised throughout the solid.

● Intermolecular forces (**between** covalent molecules)

Hydrogen bonds	–	between δ^+ H in one molecule and δ^- F, O or N in another molecule
Dispersion forces	–	between all molecules. Their strength depends upon the number of electrons in the molecule.
Dipole/dipole forces	–	between δ^+ atoms in one molecule and δ^- atoms in another molecule.

Things to learn

❑ An **ionic bond** is the electrostatic attraction that occurs between an atom that has lost one or several electrons – the cation – and one that has gained one or several electrons – the anion.

❑ A **covalent bond** occurs when two atoms share a pair of electrons. It results from the overlap of an orbital containing one electron belonging to one atom with an orbital that contains one electron that belongs to the other atom. The overlap can be head on, which results in a σ bond or, side by side, which results in a π bond (see Figure 1.3). A double bond is a σ and a π bond, with two pairs of electrons being shared.

❑ A **dative covalent bond** is a covalent bond formed when one of the overlapping orbitals contained two electrons and the other none.

❑ A **metallic bond** is the force of attraction between the sea of delocalised electrons and the positive ions which are arranged in a regular lattice.

❑ The **electronegativity** of an element is a measure of the attraction its atom has for a pair of electrons in a covalent bond.

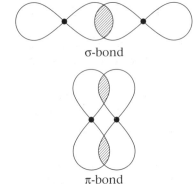

σ-bond

π-bond

Fig 1.3 σ and π bonds.

❑ A **hydrogen bond** is an intermolecular force that exists between a δ^+ hydrogen atom in one molecule and a δ^- **fluorine, oxygen or nitrogen** atom in another molecule.

❑ **van der Waals** forces are between covalent molecules and are caused by dipole/dipole and induced dipole/induced dipole (dispersion) forces.

Things to understand

Ionic bonding

- An ionic bond is likely if there is a large difference (greater than about 1.5) in the electronegativities of the two atoms.

- Cations with a **small** radius and/or high charge have a large charge density, and so are very polarising. Anions with a **large** radius and/or high charge are very polarisable. If either the cation is very polarising or the anion is very polarisable, the outer electrons in the anion will be pulled towards the cation and the bond will have some covalent character.

- Ionic bonding gives rise to an ionic lattice, which is a regular three-dimensional arrangement of ions.

- An ionic bond is, on average, the same strength as a covalent bond.

Covalent bonding

Covalent substances are either:

i   giant atomic, such as diamond, graphite and quartz (SiO_2)

ii simple molecular, such as I_2 and many organic substances

iii hydrogen-bonded molecular, such as ice, and ethanol

iv non-crystalline, such as polymers like poly(ethene).

Polar covalent bonds may result in polar molecules, but for linear molecules of formula AB_2, planar molecules of formula AB_3, tetrahedral molecules of formula AB_4 and octahedral molecules of formula AB_6, the polarities of the bonds cancel out, and the molecule is not polar.

Intermolecular forces

- The strongest intermolecular force is hydrogen bonding. In molecules with many electrons, such as I_2, the next strongest are induced dipole/induced dipole forces (sometimes called dispersion forces) and, in most cases, the weakest are permanent dipole/permanent dipole (or dipole/dipole) forces.

- The strength of dispersion forces depends mainly upon the number of electrons in the molecule. This is why I_2, with 106 electrons is a solid, whereas Cl_2 with 34 electrons is a gas.

- This explains the trend in boiling temperatures of the noble gases, as the dispersion forces are less in helium than in neon than in argon etc. Likewise the boiling temperatures of the Group 4 hydrides increase in the Group from CH_4 to PbH_4.

- The boiling temperatures of the hydrides of Groups 5, 6 and 7 (see Figure 1. 4) can also be explained. Because there is hydrogen bonding between HF molecules but not between the other hydrides in the group, HF has the highest boiling temperature. After the drop to HCl, there is a steady upward trend in boiling temperatures from HCl to HI (in spite of a decrease in the dipole/dipole forces) because the dispersion forces increase.

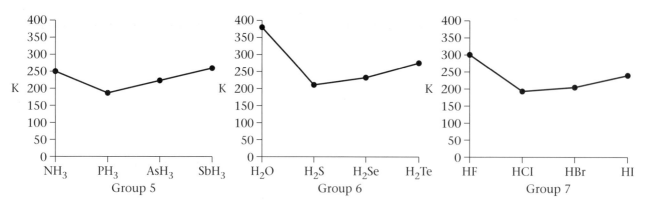

Fig 1.4 *Boiling temperatures of Groups 5, 6 and 7 hydrides*

● If the number of electrons in two different substances is about equal and neither has hydrogen bonds, then dipole/dipole forces cause a difference in boiling temperature. This is the case between butane (34 electrons, non-polar, boiling temperature – 0.5 °C) and propanone (32 electrons, but polar, so boiling temperature + 56.2 °C).

Effect of heat (melting)

When a solid is heated from room temperature until it melts:

● The particles (ions, molecules or atoms) vibrate more.

● As the temperature rises the vibrations increase until they become so great that the forces between the particles are overcome, and the regular arrangement in the lattice breaks up. The substance is then a liquid.

● In an ionic solid, such as NaCl, the vibrating particles are ions which are held by strong forces of attraction, and so the ionic solid has a high melting temperature.

● In a simple molecular solid, such as iodine (I_2) or ice, the vibrating particles are molecules which are held by weak van der Waals or hydrogen bonding forces of attraction, and so the solid has a low melting temperature.

● In a giant atomic solid, such as diamond, the vibrating particles are atoms which are held by strong covalent bonds, and so the solid has a very high melting temperature.

Shapes of molecules

● These are explained by the electron pair repulsion theory which states:

  i The electron pairs arrange themselves as far apart from each other as possible in order to minimise repulsion.

  ii The repulsion between lone pairs is greater than that between a lone pair and a bond pair, which is greater than that between two bond pairs.

● The number of σ bond pairs of electrons and lone pairs in the molecule should be counted.

● Any π bond pairs should be ignored when working out the shape of a molecule.

2 bond pairs
linear

180°

3 bond pairs
**triangular
planar**

120°

4 bond pairs
tetrahedral

$109\frac{1}{2}°$

3 bond pairs +
1 lone pair
pyramidal

<107°

2 bond pairs +
2 lone pairs
V–shaped

<105°

5 bond pairs
**Trigonal
bipyramid**

120°   90°

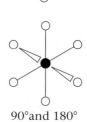

6 bond pairs
octahedral

90° and 180°

Fig 1.5 *Shapes of molecules and ions*

- The total number of pairs of electrons indicates the **arrangement** of the electrons.

> 2 = linear
> 3 = triangular planar
> 4 = tetrahedral
> 5 = trigonal bipyramid
> 6 = octahedral

But the **shape** will differ if any of the pairs are lone pairs (see Figure 1.5).

> 4 pairs: 3 bond + 1 lone:
>> molecular shape is pyramidal, e.g. NH_3, or PCl_3
> 2 bond + 2 lone:
>> molecular shape is bent or V-shaped, e.g. H_2O
> 6 pairs: 5 bond + 1 lone:
>> molecular shape is a square-based pyramid.
> 4 bond + 2 lone:
>> molecular shape is square planar.

Shapes of ions

Negative ions have gained 1 electron for each negative charge.
Thus SO_4^{2-} has 4 σ bonds (and 2 π bonds) and no lone pairs around the sulphur, and so the ion is tetrahedral.
CO_3^{2-} has 3 σ bonds (and 1 π bond) and no lone pairs around the carbon, and so it is triangular planar.
NO_3^- has one single covalent, one dative covalent and one double covalent bond (i.e. 3 σ and 1 π) and no lone pairs around the nitrogen and so is also triangular planar.
Positive ions have lost 1 electron for each positive charge.
Thus NH_4^+ has 4 σ bonds and no lone pairs around the nitrogen, and so is tetrahedral

 Checklist

Before attempting the questions on this topic, check that you understand:

- ☐ the nature of ionic, covalent and dative covalent bonds
- ☐ the effect of difference in electronegativity on type of bonding
- ☐ that polar bonds may not give rise to polar molecules
- ☐ the effect of size and charge of ions on the type of bonding
- ☐ intermolecular forces such as hydrogen bonding and van der Waals
- ☐ the different structures of solids and their properties
- ☐ the trends in boiling temperatures caused by intermolecular forces
- ☐ the nature of hydration of ions
- ☐ the changes in motion and arrangement of particles on change of state
- ☐ a metallic bond
- ☐ the shapes of molecules and ions.

Unit 1

Testing your knowledge and understanding

Answers
σ and π
Al³⁺, Mg²⁺, Na⁺, F⁻, N³⁻, O²⁻
a Al³⁺ **b** Al³⁺ **c** Al³⁺
N³⁻
K⁺
a Li⁺ **b** Li⁺
a F **b** Cl
a Ca **b** Na
a hydrogen fluoride **b** hydrogen fluoride **c** water **d** aluminium chloride
HCl, H₂O, NH₃, ICl, H₂S
Xe > Kr > Ar > Ne > He
HCl
H₂O, NH₃, HF, CH₃OH, CH₃NH₂

The answers to the numbered questions are on pages 122–123.

For the first set of questions, cover the margin, write your answers, then check to see if you are correct.

● State which types of bond are present in a double bond as in O=C=O.

● State the value of the charge on the ions of the following elements:

  Al, Mg, Na, F, N, O.

● State which of the cations above:

  a has the smallest radius
  b has the largest charge density
  c is the most polarising.

● State which of the anions above is the most polarisable.

● Which ion has the largest radius: Li⁺ or Na⁺ or K⁺?

● Which of these three ions, Li⁺, Na⁺ or K⁺, has:

  a the largest surface charge density
  b the largest polarising power?

Which is the most electronegative element in each list:
  a C, O, N, F
  b Cl, Br, I?

Which is the least electronegative element of:
  a Be, Mg, Ca
  b Na, Mg, Al?

● Which is the molecule in each group with the most polar bond:

  a HF, HCl, HBr
  b HF, H₂O, NH₃
  c H₂O, H₂S, H₂Se
  d AlCl₃, SiCl₄, PCl₅?

● List all of the following molecules that are polar:

H₂	HCl	H₂O	NH₃	CH₄
CO₂	ICl	H₂S	BH₃	CF₄

● Rearrange the noble gases in order of **decreasing** boiling temperature:

Ar, He, Kr, Ne and Xe

● State which of the following has the **lowest** boiling temperature:

HF, HCl, HBr, or HI

● Which of the following form hydrogen bonds between their molecules?

H₂O ✓	NH₃ ✓	HF ✓	HCl	CO₂	CH₄
CH₃OH ✓	CH₃NH₂ ✓	CH₃F ✓	CHCl₃	CH₂=CH₂	poly(ethene)

1 Draw a diagram of two p orbitals overlapping which produce:

  a a σ bond

  b a π bond

2 For each pair, state which is the more covalent substance and give a reason:

  a AlF₃ or AlCl₃

  b BeCl₂ or MgCl₂.

3 Intermolecular forces can be divided into three types:

    a   hydrogen bond
    b   dispersion (induced dipole/induced dipole)
    c   permanent dipole/permanent dipole.

For all the following substances state **all** the types of intermolecular forces a, b and/or c that are present:

    i HF   **ii** I_2 **iii** HBr   **iv** PH_3 **v** Ar

4 Solid structures can be divided into six types:

    a metallic, **b** giant atomic (network covalent), **c** ionic,
    d hydrogen bonded molecular, **e** simple molecular, **f** polymeric.
For each of the solids listed below, state the structural type in the solid.

iodine	ice, $H_2O(s)$	copper
silica (sand), SiO_2	dry ice, $CO_2(s)$	calcium oxide, CaO
poly(ethene)	graphite	sulphur
copper sulphate, $CuSO_4$	sucrose	lithium fluoride, LiF

5 Explain the processes of melting and boiling in terms of the arrangement and motion of the particles.

6 State, for each of the following pairs, which has the stronger forces between particles and hence has the higher boiling temperature, and explain why in terms of the types of force present:

    a NH_3 or PH_3
    b   HCl or HBr
    c CH_3COCH_3 (propanone) or C_4H_{10} (butane)
    d P_4 or S_8
    e   NaCl or CCl_4

7 In order to work out the shape of a molecule or ion, you should first evaluate the number of σ bond pairs of electrons and the number of lone (unbonded) pairs of electrons around the **central** atom.

Construct a table similar to that shown below, and use it to deduce the shapes of the following species.

SiH_4, BF_3, $BeCl_2$, PCl_3, SF_6, XeF_4, NH_4^+, PCl_6^-

Molecule/ ion	Number of σ bond pairs	Number of lone pairs	Total number of electron pairs	Shape
NH_3	3	1	4	pyramidal

8 Explain why solid sodium metal conducts electricity whereas solid sodium chloride does not.

Topic **1.4** The Periodic Table I

 Introduction

● The melting and boiling temperatures of the elements depend on the type and strength of the bond or intermolecular force between particles.

● There are trends in physical and chemical properties across a Period and down a Group. For instance

1 Elements show **decreasing** metallic character across a Period.

2 Elements show **increasing** metallic character down a Group.

 Things to learn and understand

Electronic structure

● The elements in the Periodic Table are arranged in order of atomic number, so each element has one more proton and hence one more electron than the previous element.

● Elements in the same group have the same number of electrons in their **outer** shells. These are called the valence electrons.

● Elements in the same Period have the same number of shells containing electrons so their outer or valence electrons are in the same shell.

● The order in which electrons fill the orbital types is:

1s; 2s, 2p; 3s, 3p; 4s, 3d, 4p; 5s, 4d, 5p; 6s (see Figure 1.1 on page 3)

Melting (and boiling) temperatures of the Period 3 elements (Na to Ar)

● The melting temperature depends upon the strength of the forces between particles that separate during melting (or boiling).

These particles may be:

i metal ions in a sea of electrons in metals

ii covalently bonded atoms in a giant atomic structure

iii molecules with intermolecular forces between them in simple molecular solids.

● To understand the trends in melting temperatures, you should first decide what type of bonding or force is between the particles.

● If the solid is **metallic**: the greater the charge density of the ion in the lattice, the stronger the force holding the lattice together and so the higher the melting temperature.

● If the solid is a **giant atomic** lattice: the covalent bonds throughout the lattice are strong and so the solid has a very high melting temperature.

● If the solid is a **simple molecular** substance: the melting (or boiling) temperature depends upon the strength of **dispersion** (induced dipole/induced dipole) force between the molecules.

● In Period 3, sodium, magnesium and aluminium are metallic, silicon forms a giant atomic lattice, and phosphorus, P_4, sulphur, S_8, chlorine, Cl_2 and argon, Ar, all form simple molecular solids.

Helpful hints

There are two exceptions with the first 40 elements. Because of the extra stability of a half filled or full set of d orbitals, chromium is [Ar] $4s^1$, $3d^5$, and copper is [Ar] $4s^1$, $3d^{10}$.

Helpful hints

The more electrons in the molecule, the stronger the dispersion force and the higher the melting or boiling temperature.

Electrical conductivity

Solids conduct electricity by the flow of delocalised electrons.

Thus metals conduct electricity. Graphite also conducts, but only in the plane of the layers. This is due to the π-electrons that are delocalised above and below the layers.

 Checklist

Before attempting the questions on this topic, check that you can:

☐ Write the electronic structure for elements numbers 1 to 36.

☐ Explain the variation of melting and boiling temperatures for the elements in Period 3.

☐ Explain the change in electrical conductivity of the elements in Period 3.

☐ Explain the variation in ionisation energies of the elements in Period 3.

 Testing your knowledge and understanding

 The answers to the numbered questions are on page 123.

1 What, in terms of electronic structure, are the features that the following have in common:

a members of the same group , e.g. Group 2,

b members of the same period, e.g. Period potassium to krypton?

2 State the type of solid structure of the elements listed. Give your answer as one of:

metallic, giant atomic, ionic, hydrogen bonded molecular, simple molecular, polymeric:

a hydrogen   **b** sodium   **c** silicon   **d** sulphur
e chlorine   **f** argon.

3 Explain the difference in melting temperatures of the following elements:

Element	Na	Mg	Si	P_4	S_8	Cl_2
Melting temperature/^{0}C	98	650	1410	44	113	−101

4 The ionisation energies of sodium and magnesium are listed below:

Element	1st ionisation energy/kJ mol^{-1}	2nd ionisation energy/kJ mol^{-1}
Sodium	494	4560
Magnesium	736	1450

a Explain the meaning of: **i** nuclear charge, **ii** screening (or shielding) by inner electrons.

b Use the concepts explained in (a) to explain why: **i** The 2nd ionisation energy of sodium is very much more than its 1st ionisation energy. **ii** The 1st ionisation energy of sodium is less than the 1st ionisation energy of magnesium.

Unit 1

Topic **1.5**

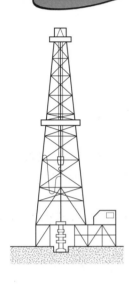

OIL RIG

Introduction to oxidation and reduction

 Introduction

● Redox reactions are those which involve a transfer of electrons
● Remember **OIL RIG** (**O**xidation **I**s **L**oss, **R**eduction **I**s **G**ain of electrons)

Things to learn

❏ **Oxidation** occurs when a substance loses one or more electrons. There is an increase in the oxidation number of the element involved.

❏ **An oxidising agent** is a substance that oxidises another substance and so is itself reduced. The half equation involving an oxidising agent has electrons on the left-hand side, i.e. it **takes** electrons from the substance being oxidised.

❏ **Reduction** occurs when a substance gains one or more electrons. There is a decrease in the oxidation number of the element involved.

❏ **A reducing agent** is a substance that reduces another substance and so is itself oxidised. The half equation involving a reducing agent has electrons on the right-hand side, i.e. it **gives** electrons to the substance being reduced.

Things to understand

Oxidation number

● The oxidation number is the charge on an atom of the element in a compound calculated assuming that all the atoms in the compound are simple monatomic ions. The more electronegative element is given an oxidation number of –1 per bond.

● There are some rules used for calculating oxidation numbers. They should be applied in the following order:

 1 The oxidation number of an uncombined element is zero.

 2 A simple monatomic ion has an oxidation number equal to its charge.

 3 The oxidation number of Group 1 metals is always +1, and of Group 2 metals is +2.

 4 Fluorine always has an oxidation number of –1, hydrogen (except in metallic hydrides) of +1, and oxygen (except in F_2O and peroxides) of –2.

 5 The sum of the oxidation numbers in a molecule adds up to 0, and those in a polyatomic ion (such as SO_4^{2-}) add up to the charge on the ion.

● When an element is oxidised, its oxidation number increases.

Helpful hints

In an overall equation, the total increase in oxidation number of one element (increase x number of atoms of that element) must equal the total decrease in oxidation number of another element.

Worked example

Calculate the oxidation number of chlorine in Cl_2, Cl^-, $MgCl_2$ and ClO_3^-.

Answers: in Cl_2 = 0 (uncombined element: rule 1)
in Cl^- = −1 (monatomic ion: rule 2)
in $MgCl_2$ = −1 (Mg is 2+; (+2) +2Cl = 0; therefore each Cl is −1: rules 3 and 5)
in ClO_3^- = +5 (Cl + 3 × (−2) = −1 : rules 4 and 5)

Half equations

These are written:

- either as reduction with electrons on the left side of the half equation,

$$Cl_2 (g) + 2e^- \rightleftharpoons 2Cl^- (aq)$$

here chlorine is being reduced and so is acting as an oxidising agent.

- or as oxidation with electrons on the right side of the equation,

$$Fe^{2+} (aq) \rightleftharpoons Fe^{3+} (aq) + e^-$$

here iron(II) ions are being oxidised, and thus are acting as a reducing agent.

- Many oxidising agents only work in acid solution. Their half equations have H^+ ions on the left hand side and H_2O on the right. This is likely with oxdising agents containing oxygen (such as MnO_4^-),

$$MnO_4^-(aq) + 8H^+(aq) + 5e^- \rightleftharpoons Mn^{2+}(aq) + 4H_2O(l)$$

- If a redox system is in alkaline solution, OH^- may need to be on one side and H_2O on the other,

$$Cr^{3+}(aq) + 8OH^-(aq) \rightleftharpoons CrO_4^{2-} + 4H_2O(l) + 3e^-$$

Overall redox equations

- Overall redox equations are obtained by adding half equations together.
- One half equation must be written as a reduction (electrons on the left) and the other as an oxidation (electrons on the right).
- When they are added the electrons must cancel. To achieve this it may be necessary to multiply one or both half equations by integers,

e.g. for the overall equation for the oxidation of Fe^{2+} ions by MnO_4^- ions

add $MnO_4^-(aq) + 8H^+(aq) + 5e^- \rightleftharpoons Mn^{2+}(aq) + 4H_2O(l)$

to 5 × $Fe^{2+}(aq) \rightleftharpoons Fe^{3+}(aq) + e^-$

$$MnO_4^-(aq) + 8H^+(aq) + 5Fe^{2+}(aq) \rightleftharpoons Mn^{2+}(aq) + 4H_2O(l) + 5Fe^{3+}(aq)$$

Helpful hints

Remember that an oxidising agent becomes reduced (loses electrons), and a reducing agent becomes oxidised.
Half equations must balance for atoms **and** for charge.
For example, both sides of the Cr^{3+} equation add up to −5.

Helpful hints

If the question asks for the overall equation for the reaction of A with B, make sure that **both** A and B appear on the left-hand side of the final overall equation.

Checklist

Before attempting the questions on this topic, check that you can:

☐ Define oxidation and reduction.

☐ Define oxidising agent and reducing agent.

☐ Calculate oxidation numbers in neutral molecules and in ions.

☐ Write ionic half equations.

☐ Combine half equations and so deduce an overall redox equation.

 Testing your knowledge and understanding

For the first set of questions, cover the margin, write your answer, then check to see if you are correct.

● In the following equations, state which substance, if any, has been oxidised:

  a $2Ce^{4+}$ (aq) + $2I^-$ (aq) ⇌ $2Ce^{3+}$ (aq) + I_2 (aq)
  b H^+ (aq) + OH^- (aq) ⇌ H_2O (l)
  c Zn (s) + $2H^+$ (aq) ⇌ Zn^{2+} (aq) + H_2 (g)
  d $2Fe^{2+}$ (aq) + $2Hg^{2+}$ (aq) ⇌ $2Fe^{3+}$ (aq) + Hg_2^{2+} (aq)

● Calculate the oxidation number of the elements in bold in the following:

  SO_2, **H_2S**,
  CrO_4^{2-},   **$Cr_2O_7^{2-}$**,
  H_2O_2,   **H_2SO_4**,   **Fe_3O_4**

Answers
a iodide ions
b none
c zinc
d iron(II) ions

+4 −2
+6 +6
−1 +6 $2\frac{2}{3}$ (1 at +2 and 2 at +3; average $2\frac{2}{3}$)

 The answers to the numbered questions are on pages 123–124.

1 Construct ionic half equations for:

  a Sn^{2+} ions being oxidised to Sn^{4+} ions in aqueous solution
  b Fe^{3+} ions being reduced to Fe^{2+} ions in aqueous solution.
  c Now write the balanced equation for the reaction between Fe^{3+} and Sn^{2+} ions in aqueous solution.

2 Construct ionic half equations for:

  a hydrogen peroxide being reduced to water in acid solution
  b sulphur being reduced to hydrogen sulphide in acid solution
  c Now write the balanced equation for the reaction between hydrogen peroxide and hydrogen sulphide in acid solution.

3 Construct ionic half equations for:

  a $PbO_2(s)$ being reduced to $PbSO_4(s)$ in the presence of $H_2SO_4(aq)$
  b $PbSO_4(s)$ being reduced to $Pb(s)$ in the presence of water
  c Now write the balanced equation for the reaction between PbO_2 and lead in the presence of dilute sulphuric acid.

Topic 1.6 Group 1 and Group 2

Introduction

❑ The properties of elements and their compounds change steadily down a Group.

❑ This means that the answer to a question about which is the most/least reactive, easiest/hardest to decompose, most/least soluble, etc. will be an element, or a compound of that element, either at the top or at the bottom of the Group.

❑ Down a Group the elements become increasingly metallic in character. Thus:
  ● their oxides become stronger bases
  ● they form positive ions more readily
  ● they form covalent bonds less readily.

 Things to learn and understand

Physical properties of the elements

● Group 1: All are solid metals; their melting temperatures and hardness **decrease** down the group; all conduct electricity.
● Group 2: All are solid metals; their melting temperatures and hardness **decrease** down the group (except magnesium which has a lower melting temperature than calcium); all conduct electricity; their melting temperatures are higher than the Group 1 element in the same period.

Flame colours of their compounds

● Group 1: lithium   carmine red
            sodium   yellow
            potassium   lilac
● Group 2: calcium   brick red
            strontium   crimson red
            barium   green.
● These colours are caused because:
  1 Heat causes the compound to vaporise and produce some atoms of the metal with electrons in a higher orbital than the ground state (e.g. in the 4th shell rather than the normal 3rd shell for sodium).
  2 The electron falls back to its normal shell and as it does so, energy in the form of visible light is emitted. The light that is emitted is of a characteristic frequency, and hence colour, dependent on the energy level difference between the two shells (see Figure 1.6).

Ionisation energies

● The value of the 1st ionisation energy for Group 1 and of the 1st and 2nd ionisation energies for Group 2 **decreases** down the group. This is

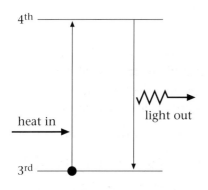

4th ⎯⎯⎯⎯⎯⎯⎯⎯⎯

heat in   light out

3rd ⎯⎯⎯⎯⎯⎯⎯⎯⎯

Fig 1.6 *Emission of a spectral line*

because as the atoms get larger, the outer electrons are further from the nucleus and so are held on less firmly. The increase in nuclear charge is compensated for by an increase in the shielding by the inner electrons.

Reactions of the elements with oxygen

> The vigour of the reaction increases down the group.

- Group 1: All burn.

    Lithium forms an oxide: $4Li + O_2 \rightarrow 2Li_2O$

    Sodium forms a peroxide: $2Na + O_2 \rightarrow Na_2O_2$

    Potassium and the others form a superoxide: $K + O_2 \rightarrow KO_2$
- Group 2: All burn to form ionic oxides of formula MO, except that in excess oxygen barium forms a peroxide (BaO_2).

    $$2Ca + O_2 \rightarrow 2CaO$$

Reaction of the elements with chlorine

- Group 1: All react vigorously to form ionic chlorides of formula MCl. These dissolve in water to produce hydrated ions, e.g.

    $$NaCl(s) \rightarrow Na^+(aq) + Cl^-(aq)$$
- Group 2: All react vigorously to produce ionic chlorides of formula MCl_2, except that $BeCl_2$ is covalent when anhydrous. All Group 2 chlorides are soluble in water producing hydrated ions of formula $[M(H_2O)_6]^{2+}$. Beryllium chloride gives an acidic solution because of deprotonation:

    $$[Be(H_2O)_6]^{2+} + H_2O \rightleftharpoons [Be(H_2O)_5(OH)]^+ + H_3O^+$$

Reaction of the elements with water

> The rate of reaction increases down the group.

- Group 1: All react vigorously with cold water to give an alkaline solution of metal hydroxide and hydrogen gas, e.g.

    $$2Na(s) + 2H_2O(l) \rightarrow 2NaOH(aq) + H_2(g)$$
- Group 2: Beryllium does not react but magnesium burns in steam to produce an **oxide** and hydrogen:

    $$Mg + H_2O \rightarrow MgO + H_2$$

    the others react rapidly with cold water to form an alkaline suspension of metal **hydroxide** and hydrogen gas:

    $$Ca + 2H_2O \rightarrow Ca(OH)_2 + H_2$$

Reactions of Group 2 oxides with water

- BeO is amphoteric and does not react with water.
- MgO is basic and reacts slowly with water to form a hydroxide.
- All the others react rapidly and exothermically to form alkaline suspensions of the hydroxide, which have a pH of about 13.

    $$CaO + H_2O \rightarrow Ca(OH)_2$$

Solubilities of Group 2 sulphates and hydroxides

- Sulphates: their solubilities **decrease** down the group. $BeSO_4$ and $MgSO_4$ are soluble; $CaSO_4$ is slightly soluble, $SrSO_4$ and $BaSO_4$ are insoluble.
- Hydroxides: their solubilities **increase** down the group. $Be(OH)_2$ and $Mg(OH)_2$ are insoluble, $Ca(OH)_2$ and $Sr(OH)_2$ are slightly soluble, and $Ba(OH)_2$ is fairly soluble.
- Addition of aqueous sodium hydroxide to solutions of Group 2 salts produces a white precipitate of metal hydroxide. (Barium produces a faint precipitate):

    $$M^{2+}(aq) + 2OH^-(aq) \rightarrow M(OH)_2(s)$$

● Addition of aqueous sulphate ions to a solution of Sr^{2+} or Ba^{2+} ions produces a white precipitate of metal sulphate:
$$Ba^{2+}(aq) + SO_4^{2-}(aq) \rightarrow BaSO_4(s)$$

Thermal stability of nitrates and carbonates

● Thermal stability **increases** down both groups:
● Group 2 nitrates all decompose to give a metal oxide, brown fumes of nitrogen dioxide, and oxygen:
$$2Ca(NO_3)_2 \rightarrow 2CaO + 4NO_2 + O_2$$
● Group 1 nitrates, except lithium nitrate, decompose to give a metal nitrite and oxygen:
$$2NaNO_3 \rightarrow 2NaNO_2 + O_2$$
but $\quad\quad 4LiNO_3 \rightarrow 2Li_2O + 4NO_2 + O_2$
● Group 2 carbonates all decompose (except barium carbonate which is stable to heat) to give a metal oxide and carbon dioxide:
$$CaCO_3 \rightarrow CaO + CO_2$$
● Group 1 carbonates are stable to heat except for lithium carbonate:
$$Li_2CO_3 \rightarrow Li_2O + CO_2$$

> The compound is more likely to be decomposed on heating if the cation polarises the anion. Thus Group 2 compounds (cation 2+) decompose more readily than Group 1 compounds (cation only 1+). Compounds of metals higher in a group (smaller ionic radius) decompose more easily than compounds of metals lower in the Group.

 Checklist

Before attempting the questions on this topic, check that you know:

☐ the physical properties of the Group 1 and Group 2 elements

☐ the flame colours caused by their compounds

☐ the trends in ionisation energies within a group

☐ the reactions of the elements with oxygen, chlorine and water

☐ the reactions of their oxides with water

☐ the oxidation states of the elements in Group 1 and in Group 2

☐ the trends in solubilities of Group 2 sulphates and hydroxides

☐ the reason for the trend in thermal stability of Group 1 and Group 2 nitrates and carbonates.

 Testing your knowledge and understanding

Answers
Lithium
K is +1, Ca is +2
It is extremely insoluble.

For the first set of questions, cover the margin, write your answer, then check to see if you are correct.

● Which Group 1 metal has the highest melting temperature?

● What is the oxidation number of:
 i potassium in $K_2Cr_2O_7$ and
 ii calcium in $CaCO_3$?

● Barium compounds normally are poisonous, but barium sulphate is given to people in order to outline their gut in radiography (X-ray imaging). Why is barium sulphate not poisonous?

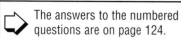

The answers to the numbered questions are on page 124.

1 You are given solid samples of three chlorides. One is lithium chloride, one is potassium chloride and the other is barium chloride. Describe the tests that you would do to find out which was which.

2 Explain why sodium compounds give a yellow colour in a flame.

3 Explain why the 1st ionisation energy of sodium is larger than the 1st ionisation energy of potassium.

4 Write balanced equations for the reactions of:
  a   calcium with oxygen
  b   calcium with water
  c   potassium with water
  d   magnesium with steam.

5 Explain why the addition of dilute sodium hydroxide to a solution of magnesium chloride produces a white precipitate, but little or no precipitate when dilute sodium hydroxide is added to a solution of barium chloride.

6 State and explain which Group 2 element forms the least thermally stable carbonate.

7 Write balanced equations for the thermal decomposition of the following, but if there is no reaction at laboratory temperatures, say so:
  a   lithium nitrate, sodium nitrate and magnesium nitrate
  b   sodium carbonate, magnesium carbonate and barium carbonate.

Topic 1.7 Group 7 (chlorine to iodine)

 Introduction

Group 7 characteristics are that:

● The halogens are oxidising agents with their strength decreasing down the group.

1 Chlorine is the strongest oxidising agent and iodine is the weakest.

2 Chloride ions are the weakest reducing agents, and so are the hardest to oxidise, whereas iodide ions are the strongest reducing agents, and so are the easiest to oxidise.

● They form halide ions, such as Cl^-, and oxoanions, such as ClO_3^-.

● Solutions of the hydrogen halides are strong acids.

 Things to learn and understand

Disproportionation

This occurs when an element is simultaneously oxidised and reduced. It follows that there must be at least two atoms of that element, with the **same** oxidation number, on the left of the equation, and that the element must be able to exist in at least three different oxidation states.

Chlorine disproportionates in alkali:

$$Cl_2 + 2OH^- \rightarrow Cl^- + OCl^- + H_2O$$
$$(0) \qquad\qquad (-1) \quad (+1)$$

Physical properties (at room temperature)

● Chlorine is a greenish gas.

● Bromine a brown liquid.

● Iodine is a dark grey lustrous solid which gives a violet vapour on heating.

● A solution of iodine in aqueous potassium iodide is red/brown, and in non-oxygen-containing organic solvents it is violet.

Tests for the elements (only examined at AS in Unit Test 3B)

● First observe the colour of the halogen or its solution.

● Chlorine rapidly bleaches damp litmus paper. It will displace bromine from a solution of potassium bromide (then test for bromine by adding a suitable organic solvent such as hexane, which will show the brown bromine colour).

● Bromine is brown and will (slowly) bleach litmus. It will displace iodine from a solution of potassium iodide (then test for iodine with starch or hexane as below).

● Iodine turns a solution of starch blue-black, and it forms a violet solution when dissolved in suitable organic solvents such as hexane.

Halides

- All hydrogen halides are covalent gases, but are soluble in water because they react with water to form ions. Their solutions are strongly acidic:
$$HCl + H_2O \rightarrow H_3O^+ (aq) + Cl^-(aq)$$
Hydrogen fluoride is a weak acid and is only partially ionised.
- All metal halides are soluble in water, except for silver and lead halides.

Test for halides (only examined at AS in Unit Test 3B)

To a **solution** of the halide add dilute nitric acid to prevent carbonates from interfering with the test. Then add silver nitrate solution followed by ammonia solution.

	Chloride	Bromide	Iodide
Addition of $Ag^+(aq)$	white precipitate	cream precipitate	yellow precipitate
Addition of dilute NH_3	precipitate dissolves	no change	no change
Addition of concentrated NH_3	precipitate dissolves	precipitate dissolves	no change

Addition of concentrated sulphuric acid to the solid halide

- Chlorides produce steamy acid fumes of HCl.
- Bromides produce steamy acid fumes of HBr with some brown bromine and some SO_2 gas.
- Iodides give clouds of violet iodine vapour.

This is due to the fact that HBr is just powerful enough as a reducing agent to reduce some of the concentrated sulphuric acid to sulphur dioxide and itself be oxidised to bromine.

The HI initially produced is a very powerful reducing agent. It reduces the concentrated sulphuric acid and is itself oxidised to iodine.

Oxidation numbers

- Chlorine is 0 in Cl_2.
- Chlorine is −1 in chlorides.
- Chlorine is +1 in ClO^- ions.
- Chlorine is +5 in ClO_3^- ions.

Redox

Chlorine is a stronger oxidising agent than bromine, which is stronger than iodine.

- Chlorine is a powerful oxidising agent and is reduced to the −1 state.
- The half equation is:
$$Cl_2(aq) + 2e^- \rightleftharpoons 2Cl^-(aq)$$
and is similar for the other halogens.
- Chlorine disproportionates in alkali at room temperature:
$$Cl_2(aq) + 2OH^-(aq) \rightarrow Cl^-(aq) + OCl^-(aq) + H_2O(l)$$
$$(0) \qquad\qquad (-1) \quad (+1)$$
- Chlorate(I) compounds disproportionate when heated:
$$3OCl^- \rightarrow 2Cl^- + ClO_3^-$$
$$3\times(+1) \quad 2\times(-1) \quad (+5)$$
- Bromine is extracted from seawater by bubbling in chlorine gas which oxidises the Br^- ions.

Checklist

Before attempting questions on this topic, check that you can recall:

☐ the physical properties of the elements

☐ the tests for the elements and for the halides

☐ the reactions of concentrated sulphuric acid with the halides

☐ examples of –1, 0, +1 and +5 oxidation states of chlorine and the disproportionation of chlorine and chlorate(I) ions

☐ the relative strengths of the halogens as oxidising agents.

 Testing your knowledge and understanding

Answers
Bromine
Add starch. The solution goes blue-black.
Chlorine
Hydrogen iodide

For the first set of questions, cover the margin, write your answer, then check to see if you are correct.

● Name the halogen that is a liquid at room temperature.

● Describe the test for iodine.

● Which of Cl_2, Br_2 and I_2 is the strongest oxidising agent?

● Which of HCl, HBr and HI is the strongest reducing agent?

 The answers to the numbered questions are on pages 124–125.

1 Why is hydrogen chloride gas very soluble in water?

2 It is thought that a sample of solid sodium carbonate has been contaminated by some sodium chloride. How would you test for the presence of chloride ions in this sample?

3 Explain why hydrogen chloride gas is produced but chlorine is not when concentrated sulphuric acid is added to solid sodium chloride, whereas iodine and almost no hydrogen iodide is produced when concentrated sulphuric acid is added to solid sodium iodide.

4 Define disproportionation and give an example of a disproportionation reaction of chlorine or a chlorine compound.

Practice Test: Unit 1

Time allowed 1hr

All questions are taken from parts of previous Edexcel Advanced GCE questions.

The answers are on pages 125–126.

1 a Complete and balance the following equations.
   i $Ca + O_2 \quad \rightarrow$ [1]
   ii $Na_2O + H_2O \quad \rightarrow$ [1]
   iii $Na_2O + HCl \quad \rightarrow$ [2]
  b State and explain the trend in thermal stability of of the carbonates of the Group 2 elements as the group is descended. [3]

(Total 7 marks)
[*May 2002 Unit Test 1 question 2*]

2 a State the **structure** of, and the type of **bonding** in, the following substances. Draw labelled **diagrams** to illustrate your answers.
   i Graphite [4]
   ii Sodium chloride [3]
  b Explain why **both** graphite and sodium chloride have high melting temperatures. [3]
  c **i** Explain why graphite is able to conduct electricity in the solid state. [2]
    ii Explain why sodium chloride conducts electricity in the liquid state. [1]

(Total 13 marks)
[*May 2002 Unit Test 1 question 6*]

3 a **i** Define the terms **atomic number** and **mass number** [2]
    ii Identify the particle that contains 35 protons, 44 neutrons and 34 electrons. [1]
    Bromine consists of two isotopes of mass numbers 79 and 81. A sample of bromine gas, Br_2, was examined in a mass spectrometer. The mass spectrum showing the molecular ions is given below.

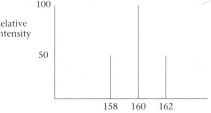

  b **i** Explain how the molecules in the sample are ionised. [2]
    ii State how the resulting ions are accelerated. [1]
    iii Identify the species responsible for the peak at $m/e = 160$. [1]

(Total 7 marks)
[*January 2001 CH1 question 1 & June 2001 Unit Test 1 question 3*]

4 Hydrogen forms compounds with most non-metallic elements and with some metals.
  a Calculate the empirical formula of the compound used in the manufacture of artificial rubber, which has the following composition by mass:
   hydrogen   11.1%   carbon   88.9% [3]
  b The boiling temperatures of hydrogen chloride and hydrogen iodide are:
   hydrogen chloride   –85 °C   Hydrogen iodide   –35 °C
   Explain why hydrogen iodide has a higher boiling temperature than hydrogen chloride [2]
  c Draw and explain the shapes of:
   i the PH_3 molecule [2]
   ii the AlH_4^- ion. [2]

d Calculate the number of **molecules** in 8.0 cm³ of gaseous phosphine, PH_3 at room temperature and pressure.
(The molar volume of a gas at room temperature and pressure should be taken as 2.4×10^4 cm³ mol⁻¹. The Avogadro constant is 6.0×10^{23} mol⁻¹.)   **[2]**
(Total 11 marks)
[January 2001 CH1 question 2]

5 a Hydrogen sulphide is produced when concentrated sulphuric acid is added to solid sodium iodide, but sulphur dioxide is produced when concentrated sulphuric acid is added to solid sodium bromide.
  i   Complete the following table:

Compound	Formula	Oxidation state of sulphur in the compound
Sulphuric acid	H_2SO_4	
Hydrogen sulphide	H_2S	
Sulphur dioxide	SO_2	

[3]

  ii Use your answers to **a** part **i** to suggest which of the ions, iodide or bromide, has the greater reducing power.   **[2]**
 b i Write an ionic half-equation to show the oxidation of chloride ions, Cl⁻, to chlorine, Cl_2.   **[1]**
  ii Write an ionic half-equation to show the reduction of chlorate(I) ions, OCl⁻, to chloride ions, in acidic solution.   **[2]**
  iii Bleach is a solution of chlorate(I) ions and chloride ions. Combine the two ionic half-equations above to produce an equation which shows the effect of adding acid to bleach.   **[1]**
(Total 9 marks)
[May 2002 Unit Test 1 question 7]

6 a Complete the following for the element sulphur
   Number of electrons in the 1st shell   =
   Number of electrons in the 2nd shell   =
   Number of electrons in the 3rd shell   =   **[1]**
 b Write one equation in each case to represent the change occurring when the following quantities are measured.
  i   The first electron affinity of sulphur   **[2]**
  ii The first ionisation of sulphur.   **[1]**
 c The graph below shows the logarithm of the successive ionisation energies of sulphur plotted against the number of the electron removed. Account for the form of the graph in terms of the electronic structure of sulphur.

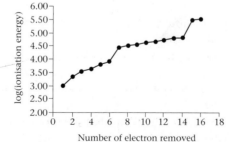

[3]

 d One of the chlorides of sulphur is sulphur dichloride, SCl_2. It is a liquid at room temperature; the electronegativity of S is 2.5 and that of chlorine is 3.0
  i   Draw a dot and cross diagram to show the bonding in SCl_2.   **[2]**
  ii The shape of the molecule is bent (V-shaped). Explain why the molecule has this shape.   **[2]**
  iii State, with reasons, whether SCl_2 is a polar molecule overall.   **[2]**
(Total 13 marks)
[January 2002 CH1 question 2 & May 2002 Unit Test 1 question 4]

2 Organic chemistry, energetics, kinetics and equilibrium

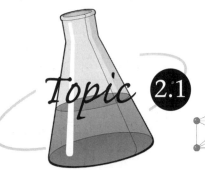

Topic 2.1 Energetics I

Introduction

- State symbols should always be used in equations in this topic.
- In definitions in this topic, there are three points to note:
  1. the chemical change that is taking place
  2. the conditions
  3. the amount of substance (reactant or product).
- It always helps to add an equation with state symbols as an example, because this may gain marks lost by omission in the word definition.
- It is assumed for simplicity that the value of ΔH is an indicator of the direction of a chemical reaction. The more exothermic the reaction, the more likely it is to take place, but other factors, such as activation energy (see page 44), also determine whether the reaction happens.

Things to learn

☐ Standard conditions are:
- gases at a pressure of 1 atmosphere
- a stated temperature (usually 298 K)
- solutions, if any, at a concentration of 1.00 mol dm^{-3}
- substances in their most stable states, e.g. carbon as graphite not diamond.

☐ Standard enthalpy of **formation**, $\Delta H_f^{\ominus}$ is the enthalpy change, under standard conditions, when **one mole** of a compound is formed from its **elements** in their standard states, e.g. $\Delta H_f^{\ominus}$ for ethanol is the enthalpy change for the reaction:
$$2C(\text{graphite}) + 3H_2(g) + \tfrac{1}{2}O_2(g) \rightarrow C_2H_5OH(l)$$

☐ Standard enthalpy of **combustion**, $\Delta H_c^{\ominus}$ is the enthalpy change, under standard conditions, when **one mole** of a substance is completely burned in **oxygen**, e.g. $\Delta H_c^{\ominus}$ for ethanol is the enthalpy change for the reaction:
$$C_2H_5OH(l) + 3O_2(g) \rightarrow 2CO_2(g) + 3H_2O(l)$$

☐ Standard enthalpy of **neutralisation** $\Delta H_{neut}^{\ominus}$ of an acid is the enthalpy change, under standard conditions, when the acid is neutralised by base and **one mole of water** is produced, e.g. $\Delta H_{neut}^{\ominus}$ of sulphuric acid, by sodium hydroxide solution, is the enthalpy change for the reaction:
$$\tfrac{1}{2}H_2SO_4(aq) + NaOH(aq) \rightarrow \tfrac{1}{2}Na_2SO_4(aq) + H_2O(l)$$

It follows from this definition that the enthalpy of formation of an element, in its standard state, is zero.

Helpful hints

You should always give an equation as an example for this and other enthalpy definitions.

for hydrochloric acid the enthalpy change is for the reaction:

$$HCl(aq) + NaOH(aq) \rightarrow NaCl(aq) + H_2O(l)$$

for any strong acid being neutralised by any strong base, the enthalpy change is for the reaction:

$$H^+(aq) + OH^-(aq) \rightarrow H_2O(l) \qquad \Delta H^\ominus = -57 \text{ kJ mol}^{-1}$$

☐ **Average bond enthalpy**. This is the **average** quantity of energy required to break **one mole** of covalent bonds in a **gaseous species** (at one atmosphere pressure). **It is always endothermic**, e.g. the C–H bond enthalpy in a gaseous compound is the enthalpy change for the reaction:

$$C–H(\text{in compound})(g) \rightarrow C(g) + H(g)$$

☐ An **exothermic** reaction gets hot, so that heat is then given out to the surroundings. **For all exothermic reactions ΔH is negative** (see Figure 2.1). This means that chemical energy is being converted into thermal (heat) energy.

☐ An **endothermic** reaction gets cold, so that heat is then taken in from the surroundings. **For all endothermic reactions ΔH is positive** (see Figure 2.2).

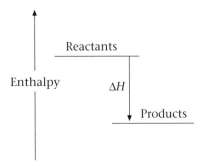

Fig 2.1 Enthalpy diagram for an exothermic reaction

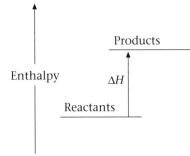

Fig 2.2 Enthalpy diagram for an endothermic reaction

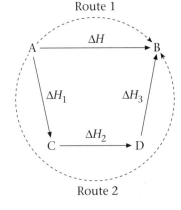

Route 1

Route 2

Fig 2.3 A reaction cycle

☐ **Hess's Law**. The enthalpy change for a given reaction is independent of the route by which the reaction is achieved (see Figure 2.3). Thus the enthalpy change proceeding **directly** from reactants to products is the same as the **sum** of the enthalpy changes of all the reactions when the change is carried out in two or more steps.

$$\Delta H = \Delta H_1 + \Delta H_2 + \Delta H_3$$

Things to understand

Calculation of $\Delta H_{\text{reaction}}$ from ΔH_f data
You can use the expression:

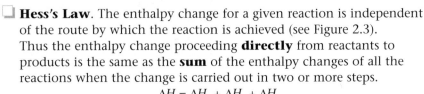

$\Delta H_{\text{reaction}}$ = the sum of $\Delta H_{\text{formation}}$ of products – the sum of $\Delta H_{\text{formation}}$ of reactants

● Remember that if you have two moles of a substance, you must double the value of its ΔH_f.

● The enthalpy of formation of an element (in its standard state) is zero.

Worked example

Calculate the standard enthalpy change for the reaction:
$$4NH_3(g) + 5O_2(g) \rightarrow 4NO(g) + 6H_2O(g)$$
given the following $\Delta H_f^{\ominus}$/kJ mol^{-1}:
$$NH_3(g) = -46.2, \quad NO(g) = +90.4, \quad H_2O(g) = -242.$$

Answer. $\Delta H_{reaction}^{\ominus}$ = sum $\Delta H_f^{\ominus}$ of products − sum $\Delta H_f^{\ominus}$ of reactants
$$= 4 \times (+90.4) + 6 \times (-242) - 4 \times (-46.2) - 5 \times (0) = -906 \text{ kJ mol}^{-1}$$

Calculation of ΔH_f from ΔH_c data

To do this you use the alternative route:

$\Delta H_2 = -\Delta H_c$ because ΔH_2 represents the reverse of combustion.
If an equation is reversed, the sign of its ΔH must be changed.

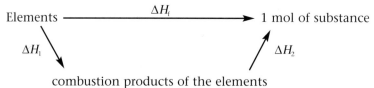

$$\text{Elements} \xrightarrow{\Delta H_f} \text{1 mol of substance}$$

ΔH_1 ↘ ↗ ΔH_2

combustion products of the elements

where ΔH_1 = the sum of the enthalpies of combustion of the elements taking into account the number of moles of each element,
and ΔH_2 = − the enthalpy of combustion of 1 mol of the substance.

Therefore: $\Delta H_{formation} = \Delta H_1 + \Delta H_2$

Worked example

Calculate the standard enthalpy of formation of propan-1-ol, $C_2H_5CH_2OH$, given the following $\Delta H^{\ominus}_{combustion}$/kJ mol^{-1}:
propan-1-ol(l) = −2010, C(graphite) = −394, H_2(g) = −286

Answer.
$$3C(s) + 4H_2(g) + \tfrac{1}{2}O_2(g) \rightarrow C_2H_5CH_2OH(l)$$

$3 \times \Delta H_c$ (graphite) ↓   ↓ $4 \times \Delta H_c(H_2)$ $-\Delta H_c(C_2H_5CH_2OH)$ ↗

$$3CO_2(g) + 4H_2O(l)$$

$\Delta H^{\ominus}_f = \Delta H^{\ominus}_c$ of elements − $\Delta H^{\ominus}_c$ of compound
$$= 3 \times (-394) + 4 \times (-286) - (-2010) = -316 \text{ kJ mol}^{-1}$$

The mass in this expression is the **total** mass of the system (usually water) not the mass of the reactants.

Calculation of ΔH values from laboratory data (only examined at AS in unit test 3B)

● Calculation of heat change from temperature change:

heat produced = mass × specific heat capacity × rise in temperature

● Then heat produced per mole = $\dfrac{\text{heat produced in experiment}}{\text{the number of moles reacted}}$

● Then enthalpy change, ΔH, is minus the heat produced per mole (for an exothermic reaction ΔH is a negative number).

● For an endothermic reaction, calculate the heat lost per mole. This equals ΔH, a positive number.

Worked example

A 0.120 g sample of ethanol was burnt and the heat produced warmed 250 g of water from 17.30 °C to 20.72 °C. The specific heat capacity of water is 4.18 J g^{-1} °C^{-1}. Calculate $\Delta H_{combustion}$ of ethanol.

Answer: Heat produced = 250 g × 4.18 J g^{-1} °C^{-1} × 3.42 °C = 3574 J.
Amount of ethanol = 0.120 g/46 g mol^{-1} = 2.61 × 10^{-3} mol
Heat produced per mole = 3574 J/2.61 × 10^{-3} mol = 1370 × 10^{3} J mol^{-1}.
$\Delta H_{combustion}$ = −1370 kJ mol^{-1}

Calculation of ΔH values from average bond enthalpies

- Draw the structural formulae for all the reactants and all the products.
- Decide what bonds are broken in the reaction and calculate the energy required to break the bonds (endothermic).
- Decide what bonds are made in the reaction and calculate the energy released by making the bonds (exothermic).
- Add the positive bond breaking enthalpy to the negative bond making enthalpy.

Worked example

Calculate the enthalpy of the reaction:

Given the following average bond enthalpies/kJ mol^{-1}:
C–C is +348, C=C is +612, H–H is +436, and C–H is +412.

Answer:
break (endo)		make (exo)	
C=C	+612	C–C	−348
H–H	+436	2 × C–H	−824
	+1048		−1172

$\Delta H_{reaction}$ = +1048 − 1172 = −124 kJ mol^{-1}

Checklist

Before attempting questions on this topic, check that you:

- ☐ Know the standard conditions for enthalpy changes.
- ☐ Know the signs of ΔH for exothermic and endothermic reactions.
- ☐ Can define the standard enthalpies of formation, combustion and neutralisation.
- ☐ Understand an energy level diagram.
- ☐ Can define and use Hess's Law.
- ☐ Know how to calculate values of ΔH from laboratory data.
- ☐ Can use bond enthalpies to calculate enthalpies of reaction.

➡ The answers to the numbered
questions are on pages 126–127.

Unit 2.

Testing your knowledge and understanding

1 State the conditions used when measuring **standard** enthalpy changes.

2 Give equations, with state symbols, which represent the following enthalpy changes :

   i   the enthalpy of formation of ethanoic acid, $CH_3COOH(l)$

   ii   the enthalpy of combustion of ethanoic acid

   iii   the enthalpy of neutralisation of an aqueous solution of ethanoic acid with aqueous sodium hydroxide.

3 Draw an enthalpy level diagram for the following sequence of reactions:

$$\tfrac{1}{2}N_2(g) + O_2(g) \rightarrow NO(g) + \tfrac{1}{2}O_2 \qquad \Delta H = +90.3 \text{ kJ mol}^{-1}$$

$$NO(g) + \tfrac{1}{2}O_2(g) \rightarrow NO_2(g) \qquad \Delta H = -57.1 \text{ kJ mol}^{-1}$$

and hence calculate the enthalpy change for the reaction:

$$\tfrac{1}{2}N_2(g) + O_2(g) \rightarrow NO_2(g)$$

4 Given that the standard enthalpies of formation of $NO_2(g)$ and $N_2O_4(g)$ are +33.9 kJ mol^{-1} and +9.7 kJ mol^{-1} respectively, calculate the enthalpy change for the reaction:

$$2NO_2(g) \rightarrow N_2O_4(g)$$

5 The standard enthalpy of combustion of lauric acid, $CH_3(CH_2)_{10}COOH(s)$, which is found in some animal fats, is –7377 kJ mol^{-1}. The standard enthalpies of combustion of C(s) and $H_2(g)$ are –394 kJ mol^{-1} and –286 kJ mol^{-1}.

Calculate the standard enthalpy of formation of lauric acid.

6 100 cm^3 of 1.00 mol dm^{-3} HCl was added to 100 cm^3 of 1.00 mol dm^{-3} NaOH in a polystyrene cup, both solutions being initially at 21.10 °C. On mixing the temperature rose to 27.90 °C. Determine the enthalpy of neutralisation and state whether the reaction is exothermic or endothermic. You may assume that the polystyrene cup has a negligible heat capacity, the solution has a density of 1.00 g cm^{-3} and that the final solution has a specific heat capacity of 4.18 J g^{-1} °C^{-1}.

7 Calculate the enthalpy change for the reaction:

$$C_2H_4(g) + H_2O(g) \rightarrow CH_3CH_2OH(g)$$

given the following average bond enthalpies in kJ mol^{-1}:

    C–C +348; C=C +612; C–H +412; H–O +463; C–O +360.

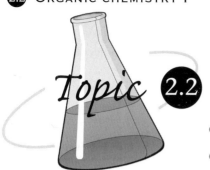

 2.2

Organic chemistry I

Introduction

☐ Organic compounds consist of a chain of one or more carbon atoms and contain functional groups (see Table 2.1). The functional group gives the compound certain chemical properties. For instance the C=C group (except in benzene rings) reacts in a similar way in all compounds. Thus knowledge of the chemistry of ethene, $H_2C=CH_2$, enables you to predict the reactions of all compounds containing the C=C group.

Substance	Alcohol	Aldehyde	Ketone	Acid	Ester	Acid chloride *	Amide *	Nitrile	Amine
Group	$-\overset{\mid}{\underset{\mid}{C}}-OH$	$\overset{H}{\underset{R}{>}}C=O$	$\overset{R}{\underset{R'}{>}}C=O$	$-C\overset{=O}{\underset{OH}{}}$	$R-C\overset{=O}{\underset{O-R'}{}}$	$-C\overset{=O}{\underset{Cl}{}}$	$-C\overset{=O}{\underset{NH_2}{}}$	$-C\equiv N$	$-\overset{\mid}{\underset{\mid}{C}}-NH_2$

Table 2.1 *Functional groups. *These will only be met at A2. R is a carbon-containing group.*

☐ You must learn the equations and conditions for the reactions in the specification.

 ## Things to learn

● **Homologous series**: a series of compounds with the same functional group, the same general formula, and where one member differs from the next by CH_2.

● **Empirical formula**: shows the simplest whole number ratio of the atoms present in one molecule.

● **Molecular formula**: shows the actual number of atoms present in one molecule, e.g. propane is C_3H_8.

● **Structural formula**: shows each atom in the molecule separately and how it is bonded. A more condensed way is to show the groupings around each carbon atom, e.g. propane can be written as $CH_3CH_2CH_3$ and propan-2-ol as $CH_3CH(OH)CH_3$.

☐ **Isomers**: two or more compounds with the same molecular formula.

☐ **Homolytic fission**: when a bond breaks with one electron going to each atom (forming radicals).

☐ **Heterolytic fission**: when a bond breaks with the two electrons going to one atom.

☐ **Substitution**: a reaction in which an atom or group of atoms in one molecule is replaced by another atom or group of atoms.

☐ **Addition**: a reaction in which two molecules react together to form a single product.

Helpful hints

When you write structural formulae, check that:
● every carbon atom has four bonds
● every oxygen has two bonds and
● every hydrogen and every halogen have only one bond.

☐ **Elimination**: a reaction in which the elements of a simple molecule such as HBr, H_2O etc. are removed from the organic molecule and not replaced by any other atom or group of atoms.

☐ **Hydrolysis**: a reaction in which water (often catalysed by aqueous acid or aqueous alkali) splits an organic molecule into two compounds.

☐ **Free radical**: a species which has a single unpaired electron, e.g. $Cl\cdot$

☐ **Nucleophile**: a species which seeks out positive centres, and must have a lone pair of electrons which it donates to form a new covalent bond.

☐ **Electrophile**: a species which seeks out negative centres, and accepts a lone pair of electrons to form a new covalent bond.

 Things to understand

Isomerism

> There are three structural isomers of formula C_3H_6O:
> ● propan-1-ol $CH_3CH_2CH_2OH$
> ● propan-2-ol $CH_3CH(OH)CH_3$
> ● methoxyethane $CH_3OCH_2CH_3$

● **Structural isomers** may have:

 i different carbon chains (straight or branched)

 ii the functional group in different places in the carbon chain

 iii different functional groups.

● **Geometric isomerism**

 i Geometric isomerism occurs when there is a C=C in the molecule, and both carbon atoms have two different atoms or groups attached.

 ii Geometric isomers are not easily interconverted because of the difficulty of rotating about the π bond.

cis

$$
\begin{array}{ccc}
CH_3 & & CH_3 \\
\diagdown & & \diagup \\
& C=C & \\
\diagup & & \diagdown \\
H & & H
\end{array}
$$

trans

$$
\begin{array}{ccc}
CH_3 & & H \\
\diagdown & & \diagup \\
& C=C & \\
\diagup & & \diagdown \\
H & & CH_3
\end{array}
$$

Fig 2.4 *Geometric isomers of but-2-ene*

Alkanes (e.g. methane, CH_4)

● Alkanes burn in excess **oxygen** to form carbon dioxide and water:

$$CH_4 + 2O_2 \rightarrow CO_2 + 2H_2O$$

● Alkanes react with **chlorine** (or bromine) in the presence of UV light in a stepwise substitution reaction:

$$CH_4 + Cl_2 \rightarrow CH_3Cl + HCl$$
$$CH_3Cl + Cl_2 \rightarrow CH_2Cl_2 + HCl$$

Alkenes (e.g. propene, $CH_3CH=CH_2$)

> The decolourisation of a brown solution of bromine is a test for a C=C group in a molecule.

● Alkenes add **hydrogen**:

$$CH_3CH=CH_2 + H_2 \rightarrow CH_3CH_2CH_3$$

conditions: pass gases over a heated nickel catalyst

● Alkenes add **bromine** (or chlorine):

$$CH_3CH=CH_2 + Br_2 \rightarrow CH_3CHBrCH_2Br$$

conditions: at room temperature with the halogen dissolved in hexane.

> With unsymmetrical alkenes, the hydrogen goes to the carbon that already has more hydrogen atoms directly bonded to it. This is Markovnikov's rule.

● Alkenes add **hydrogen halides**, e.g. HI

$$CH_3CH=CH_2 + HI \rightarrow CH_3CHICH_3$$

conditions: mix gases at room temperature.

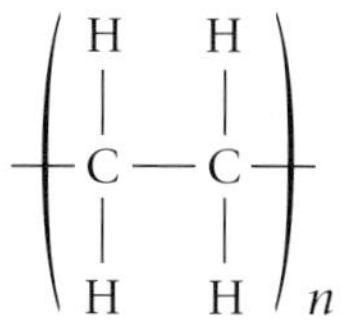

Fig 2.5
Poly(ethene)

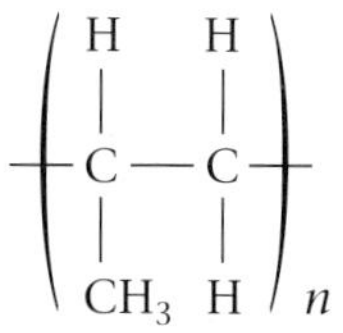

Fig 2.6
Poly(propene)

- Alkenes are **oxidised** by potassium manganate(VII) solution:

$$CH_3CH=CH_2 + [O] + H_2O \rightarrow CH_3CH(OH)CH_2OH$$

conditions: at room temperature, when mixed with sodium hydroxide solution it produces a brown precipitate.

- Alkenes can be **polymerised**. Ethene forms poly(ethene) (Figure 2.5) and propene forms poly(propene) (Figure 2.6):

$$n\text{-}CH_2=CH_2 \rightarrow -(CH_2-CH_2)-_n$$

conditions either: a very high pressure (about 2000 atm) and a temperature of about 250 °C

or: a catalyst of titanium(IV) chloride and triethyl aluminium at 50 °C and pressure of about 10 atm.

Halogenoalkanes (e.g. 2-iodopropane, CH_3CHICH_3)

- They **substitute** with **aqueous sodium (or potassium) hydroxide** to give an alcohol:

$$CH_3CHICH_3 + NaOH \rightarrow CH_3CH(OH)CH_3 + NaI$$

conditions: boil under reflux with aqueous sodium hydroxide.

- They **eliminate** with **ethanolic potassium hydroxide** to give an alkene:

$$CH_3CHICH_3 + KOH \rightarrow CH_3CH=CH_2 + KI + H_2O$$

conditions: boil under reflux with a concentrated solution of potassium hydroxide in ethanol, and collect the gaseous propene over water.

- They **substitute** with **potassium cyanide**:

$$CH_3CHICH_3 + KCN \rightarrow CH_3CH(CN)CH_3 + KI$$

conditions: boil under reflux with a solution of potassium cyanide in a mixture of water and ethanol.

- With **ammonia** they form amines:

$$CH_3CHICH_3 + 2NH_3 \rightarrow CH_3CH(NH_2)CH_3 + NH_4I$$

conditions: heat excess of a concentrated solution of ammonia in ethanol with the halogenoalkane in a sealed tube.

- The test for halogeno compounds:
  1 Warm the substance with aqueous sodium hydroxide.
  2 Add dilute nitric acid until the solution is just acidic.
  3 Add silver nitrate solution.

Result:
- white precipitate, soluble in dilute ammonia indicates chloro-compound
- cream precipitate, soluble in concentrated ammonia indicates bromo-compound
- yellow precipitate, insoluble in concentrated ammonia indicates iodo-compound.

The rate of all these reactions increases C–Cl to C–Br to C–I, because the bond enthalpy and hence the bond strength decreases C–Cl to C–Br to C–I. The weaker the bond the lower the activation energy and hence the faster the reaction.

Alcohols

Primary (1°) alcohols contain the CH_2OH group.

Secondary (2°) alcohols have two C atoms attached to the CHOH group.

Tertiary (3°) alcohols have three C atoms attached to the COH group (see Figure 2.7).

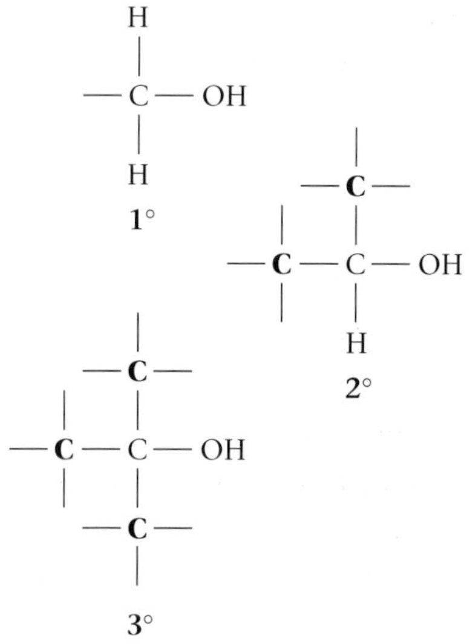

Fig 2.7 1°, 2° and 3° alcohols

Unit 2

- **Oxidation** reaction with aqueous orange **potassium dichromate(VI)**, $K_2Cr_2O_7$, acidified with dilute sulphuric acid.

  a 1° alcohols are oxidised via an aldehyde to a carboxylic acid. The solution turns green ($[Cr(H_2O)_6]^{3+}$ ions formed).

  $$CH_3CH_2OH + [O] \rightarrow CH_3CHO + H_2O$$

  $$\text{then } CH_3CHO + [O] \rightarrow CH_3COOH$$

  conditions: to stop at the **aldehyde**: add potassium dichromate(VI) in dilute sulphuric acid to the hot alcohol and allow the aldehyde to distil off.

  to prepare the **acid**: boil the alcohol with excess acidified oxidising agent under reflux.

  b 2° alcohols are oxidised to a ketone. The solution turns green.

  $$CH_3CH(OH)CH_3 + [O] \rightarrow CH_3COCH_3 + H_2O$$

  conditions: boil the alcohol and the acidified oxidising agent under reflux.

  c 3° alcohols are not oxidised and so do **not** turn the solution green.

- **Dehydration** of an alcohol to an alkene:

  $$C_2H_5OH \rightarrow H_2C{=}CH_2 + H_2O$$

  conditions: add excess concentrated sulphuric acid (or concentrated phosphoric acid) to the alcohol and heat to 170 °C.

- **Halogenation** 1°, 2° and 3° alcohols react with:

  1 Phosphorus pentachloride, PCl_5, to form a chloroalkane:

  $$C_2H_5OH + PCl_5 \rightarrow C_2H_5Cl + HCl + POCl_3$$

  conditions: at room temperature and the alcohol must be dry.

  2 Solid sodium bromide and 50% sulphuric acid to give a bromoalkane:

  $$NaBr + H_2SO_4 \rightarrow HBr + NaHSO_4$$

  $$\text{then } HBr + C_2H_5OH \rightarrow C_2H_5Br + H_2O$$

  conditions: add the sulphuric acid mixed with the alcohol to solid sodium bromide at room temperature.

  3 Phosphorus and iodine to give the iodoalkane:

  $$2P + 3I_2 \rightarrow 2PI_3$$

  $$\text{then } 3C_2H_5OH + PI_3 \rightarrow 3C_2H_5I + H_3PO_3$$

  conditions: add the alcohol to a mixture of moist red phosphorus and iodine at room temperature.

  A summary of reactions is shown in Figure 2.8.

Helpful hints

In equations in organic chemistry, an oxidising agent can be written as [O] and a reducing agent as [H], but the equation must still balance.

Note that this reaction works for all alcohols, and isomers of the alkene may be produced.

This is a test for an OH group. Steamy acidic fumes are given off when phosphorus pentachloride is added to the dry organic substance.

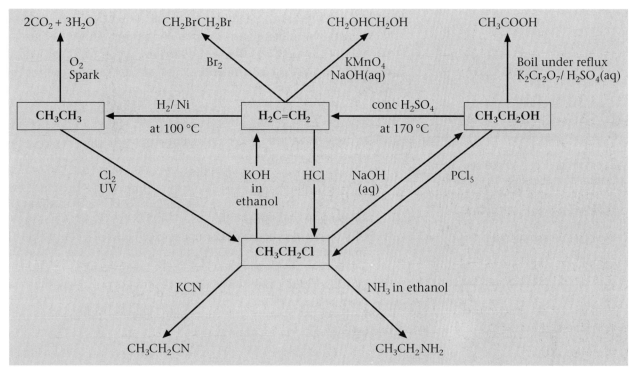

Fig 2.8 Summary of
organic reactions

Bonding and reactivity

● Bond strength is the dominant factor that determines reactivity.

a A π-bond between two carbon atoms is weaker than a σ bond between two carbon atoms. Thus alkenes are more reactive than alkanes and react by addition, whereas alkanes react by substitution.

b A C–I bond is weaker than a C–Cl bond, and so iodoalkanes are rapidly hydrolysed by aqueous sodium hydroxide, unlike chloroalkanes which react very slowly.

Quantitative organic chemistry

● This involves calculation of empirical formulae from percent data (see page 6).

● This involves calculation of percentage yield.

First calculate the theoretical yield from the equation (see page 8),

then the % yield = $\dfrac{\text{actual yield in grams}}{\text{theoretical yield in grams}} \times 100\%$

The percent yield is less than 100% because of competing reactions and handling losses.

Applied organic chemistry

● **Liquid versus gaseous fuels**. You should consider the following points:

1 the ease with which gaseous fuels can be piped into the home

2 the easier handling of liquid fuels at a filling station for cars

3 the extent and type of pollution produced:

a the quantity of carbon dioxide and any other greenhouse gases produced per kilojoule of energy

b the emission of oxides of nitrogen and sulphur, and the way in which these can be limited

c the emission of particulates

4 the energy produced per unit mass of fuel for aeroplanes

5 the extent to which the world's supply of fossil fuels is limited.

- **Polymers**. You should realise that simple polyalkenes are resistant to breaking down under environmental conditions. This is because of the strength of the C–C and the C–H bonds.
- **Organic halogen compounds**. There are three main uses of these:
  1. As polymers. PVC is used as a waterproofing material, as an electrical insulator and as a stable and maintenance free material for window frames. When burnt it produces harmful fumes of hydrogen chloride. PTFE is used as a non-stick coating for saucepans etc. These polymers are very stable owing to the strength of carbon–halogen and C–C bonds.
  2. As herbicides. Complicated chlorine compounds, such as 2,4-D and 2,4,5-T are used as selective weedkillers. They are stable and so persist in the environment.
  3. Chlorofluorocarbons (CFCs) make excellent refrigerants, but they are so stable in air and water that they diffuse to the stratosphere where they are broken down by light energy and form chlorine radicals. These catalyse the destruction of ozone. New substances have been developed which contain hydrogen as well as chlorine, fluorine and carbon, and these are less stable and are broken down by atmospheric oxidation.

> DDT is a chlorine-containing pesticide and it has eradicated malaria from some countries, but its overuse has led to the destruction of wildlife. It is a very inert chemical (owing to the strength of the C–Cl bond) and so persists in an organism, being stored in fatty tissue.

 Checklist

Before attempting the questions on this topic, check that you:

- ❏ Can name simple organic molecules.
- ❏ Can write structural formulae of structural and geometric isomers.
- ❏ Know the reactions of alkanes with air and halogens.
- ❏ Know the reactions of alkenes with hydrogen, halogens, hydrogen halides, potassium manganate(VII) and in polymerisation.
- ❏ Know the reactions of halogenoalkanes with potassium hydroxide (both aqueous and ethanolic), potassium cyanide and ammonia.
- ❏ Know the oxidation, dehydration and halogenation reactions of alcohols.
- ❏ Can relate reactivity to bond strength and polarity and to the stability of intermediates.
- ❏ Can calculate the empirical formula of a compound.
- ❏ Understand the advantages and disadvantages of the use of organic compounds.

 Testing your knowledge and understanding

For the first set of questions, cover the margin, write your answer, then check to see if you are correct.

Answers
ethene and but-1-ene
2-chloro-3-methylpentan-1-ol

- Which of the following compounds are members of the same homologous series:

  ethene C_2H_4, cyclopropane C_3H_6, but-1-ene $CH_3CH_2CH=CH_2$,

  but-1,3-diene $CH_2=CHCH=CH_2$, cyclohexene C_6H_{10}?
- Name $CH_3CH_2CH(CH_3)CHClCH_2OH$

Answers
tetrachloromethane, CCl$_4$
C$_2$H$_4$O

 The answers to the numbered questions are on pages 127–128.

● What product is obtained if a large excess of chlorine is mixed with methane and exposed to diffused light?

● A compound X contained 54.5% carbon, 36.4% oxygen and 9.1% hydrogen by mass. Calculate its empirical formula.

1 Write the structural formulae of:

   a 1,1-dibromo-1,2-dichloro-2,3-difluoropropane

   b 1-chlorobutan-2-ol.

2 Write out the structural formulae of the isomers of:

   a C$_3$H$_8$O (alcohols only)

   b C$_5$H$_{12}$

   c C$_4$H$_8$ (no cyclic compounds).

3 Define:

   a free radical

   b homolytic fission.

4 Define:

   a an electrophile

   b heterolytic fission.

5 Write equations and give conditions for the reaction of:

   a ethane + chlorine

   b ethane + oxygen.

6 Write equations and give conditions for the reaction of propene with:

   a hydrogen

   b bromine *

   c hydrogen iodide

   d potassium manganate(VII) solution *

   For reactions marked *, state what you would see.

7 Write equations and name the products obtained when 2-iodopropane is:

   a shaken with aqueous dilute sodium hydroxide

   b heated under reflux with a concentrated solution of potassium hydroxide in ethanol

   c heated in a sealed tube with a concentrated solution of ammonia in ethanol.

8 Write the equation and give the conditions for the reaction of propan-2-ol with sulphuric acid.

9 Write the structural formula of the organic product formed, if any, give its name and say what you would see when:

   a propan-1-ol is heated under reflux with dilute sulphuric acid and excess aqueous potassium dichromate(VI)

   b 2-methylpropan-2-ol is heated under reflux with dilute sulphuric acid and aqueous potassium dichromate(VI)

   c phosphorus pentachloride is added to propan-2-ol.

10 Why does bromine react rapidly with ethene at room temperature but only slowly with ethane?

Unit 2

11 Why does 1-bromopropane react more slowly than 1-iodopropane with water?

12 When 5.67 g of cyclohexene, C_6H_{10}, reacted with excess bromine, 15.4 g of $C_6H_{10}Br_2$ was obtained. Calculate the theoretical yield of $C_6H_{10}Br_2$ and hence the percentage yield of the reaction.

13 Explain why compounds such as CCl_2F_2 are harmful to the environment.

Topic **2.3** Kinetics I

- The rate of a reaction is determined by:
  1 the rate at which the molecules collide
  2 the fraction of the colliding molecules that possess enough kinetic energy 'to get over the activation energy barrier'
  3 the orientation of the molecules on collision.

 Things to learn

- ❏ **Activation energy** is the minimum energy that the reactant molecules must have when they collide in order for them to form product molecules.
- ❏ **Factors which control the rate of a reaction are:**
  1 the concentration of a reactant in a solution
  2 the pressure, if the reactants are gases
  3 the temperature
  4 the presence of a catalyst
  5 the surface area of any solids
  6 light for photochemical reactions.

 Things to understand

Collision theory

- The effect of an increase in **concentration** of a solution, or the increase in pressure of a gas, is to increase the **frequency of collision** of the molecules, and hence the rate of reaction.
- The effect of **heating** a gas or a solution is to make the molecules or ions:
  1 move faster and so have a greater average kinetic energy
  2 which increases the fraction of colliding molecules with a combined energy greater than or equal to the activation energy
  3 which results in a greater proportion of successful collisions.
- The effect of heating can be shown by the Maxwell–Boltzmann distribution (see Figure 2.9) of molecular energies at two temperatures T_1 and T_2 where T_2 is $> T_1$.

An increase in temperature also causes an increase in the rate of collision, but this is unimportant compared with the increase in the fraction of molecules with the necessary activation energy.

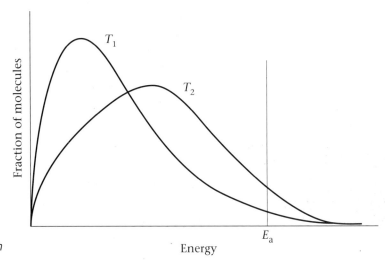

Note that the peak has moved to the right for T_2, and that the total areas under the curves are the same, but the area under the curve to the **right** of E_a equals the fraction of the molecules with energy $\geqslant E_a$, which is greater for T_2 than for T_1.

Fig 2.9 *Maxwell–Boltzmann distribution*

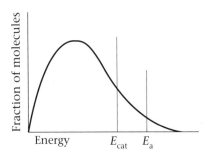

Fig 2.10 *Effect of a catalyst*

● **Catalysts** work by providing an alternative route with a lower activation energy. Thus, at a given temperature, a greater proportion of the colliding molecules will possess this lower activation energy by following the route with the catalyst, and so the reaction will be faster. This is shown on a Maxwell–Boltzmann diagram (see Figure 2.10).

● The enthalpy profile diagrams for an uncatalysed and a catalysed reaction are shown in Figure 2.11.

Do not say that a catalyst lowers the activation energy.

Note that the ΔH values are the same for both paths.

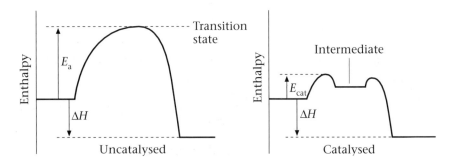

Fig 2.11 *Energy profile diagrams*

Thermodynamic stability

● This is when the enthalpy level of the products is much higher than the enthalpy level of the reactants.

● Thus the substances on the left-hand side of a very **endothermic** reaction are said to be thermodynamically stable relative to those on the right.

● The substances on the left-hand side of a very **exothermic** reaction are said to be thermodynamically unstable relative to those on the right. Whether a reaction will then take place depends upon its kinetic stability.

Kinetic stability

If a reaction has such a high value of activation energy that no molecules possess sufficient energy on collision to react, the system is said to be kinetically stable. An example is a mixture of petrol and air, which is

Unit 2

thermodynamically unstable but kinetically stable. No reaction occurs unless the mixture is ignited.

 Checklist

Before attempting the questions on this topic, check that you:

❏ Can recall the factors which effect the rate of reaction.

❏ Can explain these rate changes in terms of collision theory.

❏ Can draw the Maxwell–Boltzmann distribution of molecular energies at two different temperatures.

❏ Can use this to explain the effect of a change of temperature and the addition of catalyst.

❏ Understand the concepts of thermodynamic and kinetic stability.

 Testing your knowledge and understanding

Answers
Pressure, temperature, and catalyst
Concentration, temperature and catalyst
The surface area of the solid
Fe (Haber), or V₂O₅ (Contact), or Ni or Pt (addition of hydrogen to C=C)

 The answers to the numbered questions are on page 128.

For the following questions, cover the margin, write your answer, then check to see if you are correct.

● For a reaction involving gases, state three factors that control the rate of the reaction.

● For a reaction carried out in solution, state three factors that control the rate of the reaction.

● For the reaction of a solid with a gas or with a solution, state one other factor that controls the rate of reaction.

● Give an example of a solid catalyst used in a gas phase reaction.

1 Define

  i activation energy

  ii catalyst.

2 Explain, in terms of collision theory, why changes in temperature and in pressure and the addition of a catalyst alter the rate of a gas phase reaction.

3 Draw the Maxwell–Boltzmann distribution of energy curve for a gas:

  i at room temperature (mark this T_1) and

  ii at 50 °C (mark this T_2).

Now mark in a typical value for the activation energy of a reaction that proceeds steadily at room temperature.

4 Explain, in terms of activation energy, why animal products such as meat and milk stay fresher when refrigerated.

5 Draw energy profile diagrams of:

  i   an exothermic reaction occurring in a single step

  ii   the same reaction in the presence of a suitable catalyst

  iii a reaction where the reactants are thermodynamically stable

  iv a reaction where the reactants are both thermodynamically and kinetically unstable.

Topic 2.4 Chemical equilibria I

Introduction

Many reactions do not go to completion because the reaction is reversible. As the rate of a reaction is dependent on the concentration of the reactants, the reaction will proceed up to the point at which the rate of the forward reaction equals the rate of the reverse reaction, when there is no further change in concentrations. The system is then said to be at equilibrium.

Things to learn and understand

Dynamic equilibrium

- At equilibrium, the rate of the forward reaction equals the rate of the reverse reaction.
- Both products and reactants are constantly being made and used up, but their concentrations do not change.
- This can be demonstrated by using isotopes. For the reaction:

$$CH_3COOH(l) + C_2H_5OH(l) \rightleftharpoons CH_3COOC_2H_5(l) + H_2O(l)$$

Mix the four substances in their **equilibrium concentrations**, but have the water made from the isotope ^{18}O. After some time the ^{18}O isotope will be found (by means of a mass spectrometer) in both the ethanoic acid and the water, but the concentrations of the four substances will not have changed.

Effect of changes in conditions on position of equilibrium

- Le Chatelier's principle may help you to predict the direction of the change in the position of equilibrium, but it does not explain it.
- **Temperature.** An **increase** in temperature will move the position of equilibrium in the **endo**thermic direction. Likewise a decrease in temperature will move the equilibrium in the exothermic direction:

$$N_2(g) + 3H_2(g) \rightleftharpoons 2NH_3(g) \quad \Delta H = -92.4 \text{ kJ mol}^{-1}$$

As this reaction is exothermic left to right, an increase in temperature will cause **less** ammonia to be made, thus lowering the yield.

- **Pressure.** This applies only to reactions involving gases. An increase in pressure will drive the equilibrium to the side with fewer gas molecules. Thus, for the reaction above, an increase in pressure will result in more ammonia in the equilibrium mixture, i.e. an increased yield. This is because there are only two gas molecules on the right of the equation and four on the left.

- **Concentration.** This applies to equilibrium reactions in solution. If a substance is physically or chemically removed from an equilibrium system, the equilibrium will shift to make more of that substance:

Helpful hints

Le Chatelier's principle states that when one of the factors governing the position of equilibrium is changed, the position will alter in such a way as if to restore the original conditions.

Helpful hints

Do not say 'the side with a smaller volume', as a gas will always fill its container.

$$2CrO_4^{2-}(aq) + 2H^+(aq) \rightleftharpoons Cr_2O_7^{2-}(aq) + H_2O(l)$$

Addition of alkali will remove the H^+ ions, causing the equilibrium to move to the left.

● **Catalyst**. This has **no** effect on the position of equilibrium. What it does is to **increase the rate** of reaching equilibrium, thus a catalyst allows a reaction to be carried out at a reasonable rate at a lower temperature.

Optimum industrial conditions

Temperature

Many industrial reactions, such as the Haber process for manufacturing ammonia, are reversible and exothermic. For such reactions:

● If a high temperature is used, the yield at equilibrium is **small**, but the rate of reaction is **fast**.

● If a low temperature is used, the theoretical yield is **higher**, but the rate of reaction is **slow**.

● In systems such as this, a catalyst is used to allow the reaction to proceed rapidly at a temperature at which the yield is reasonably good. This is often called a compromise temperature, balancing yield with rate. Any unreacted gases are then separated from the products and recycled back through the catalyst chamber.

Pressure

All industrial processes that involve passing gases through a bed of catalyst must work above 1 atmosphere pressure in order to force the gases through the system.

● High pressures are extremely expensive, and are only used if the yield at lower pressures is too small to be economic.

● Two examples of manufacturing processes that use very high pressures are the Haber process and the polymerisation of ethene.

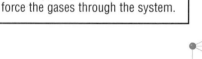 *Checklist*

Before attempting the questions on this topic, check that you:

☐ Understand that equilibria are dynamic.

☐ Can deduce the effect of changes in temperature, pressure and concentration on the position of equilibrium.

☐ Can predict the economic conditions for an industrial process.

 Testing your knowledge and understanding

 The answers to the numbered questions are on page 129.

1 Consider the reversible reaction at equilibrium:

$$A(g) \rightleftharpoons B(g)$$

Which statements are/is true about this system?

i There is no further change in the amounts of A or B.

ii No reactions occur, now that it has reached equilibrium.

iii The rate of formation of B is equal to the rate of formation of A.

2 Consider the equilibrium reaction:

$$N_2O_4(g) \rightleftharpoons 2NO_2(g) \quad \Delta H = +58.1 \text{ kJ mol}^{-1}$$

State and explain the effect on the position of equilibrium of:

i decreasing the temperature

ii halving the volume of the container

iii adding a catalyst.

3 Ethanoic acid and ethanol react reversibly:

$$CH_3COOH + C_2H_5OH \rightleftharpoons CH_3COOC_2H_5 + H_2O \quad \Delta H = 0 \text{ kJ mol}^{-1}$$

i Explain the effect of adding an alkali.

ii What will happen to the position of equilibrium if the temperature is increased from 25 °C to 35 °C?

4 White insoluble lead(II) chloride reacts reversibly with aqueous chloride ions to form a colourless solution:

$$PbCl_2(s) + 2Cl^-(aq) \rightleftharpoons PbCl_4^{2-}(aq)$$

State and explain what you would see when concentrated hydrochloric acid is added to the equilibrium mixture.

5 In the manufacture of sulphuric acid, the critical reaction is:

$$2SO_2(g) + O_2(g) \rightleftharpoons 2SO_3(g) \quad \Delta H = -196 \text{ kJ mol}^{-1}$$

The reaction is very slow at room temperature. Why are conditions of 725 K and a catalyst of vanadium(V) oxide used?

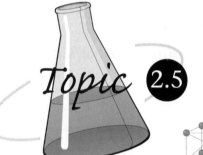

Topic **2.5** Industrial inorganic chemistry

 Introduction

Industrial chemists always need to keep manufacturing costs down. They do this by:

1 making the reaction as fast as possible

2 ensuring a high yield

3 keeping the temperature and pressure as low as possible.

 Things to learn and understand

The Haber process for the manufacture of ammonia

$$N_2(g) + 3H_2(g) \rightleftharpoons 2NH_3(g) \quad \Delta H = -92.4 \text{ kJ mol}^{-1}$$

● Because this reaction, left to right, is exothermic and the number of gas molecules decreases, the yield and kinetics are affected according to the table below:

Reaction conditions	Effect on yield	Effect on rate
Increase in temperature	Decrease	Increase
Increase in pressure	Increase	Very slight increase
Addition of catalyst	None	Very large increase

● Thus the following conditions are found to be the most economic:

Temperature: 400 °C. A higher temperature would reduce the yield, and a lower one would make the reaction uneconomically slow.

Pressure: *200 atm.* The yield at 1 atm is too low, and so a high pressure is necessary even though it is very expensive.

Catalyst: *Iron.* This allows the reaction to proceed at a fast, and hence economic, rate at a moderate temperature.

Yield per cycle: *15%.* The gases from the catalyst chamber are cooled in order to liquefy the ammonia, and then the unreacted nitrogen and hydrogen are recycled, giving a final conversion of nearly 100%.

The oxidation of ammonia to nitric acid

This happens in three stages:

1 The ammonia and air are passed over a platinum/rhodium catalyst at 900 °C:

$$4NH_3(g) + 5O_2(g) \rightleftharpoons 4NO(g) + 6H_2O(g)$$

2 On cooling, the nitrogen(II) oxide reacts with more air:

$$4NO(g) + 2O_2(g) \rightleftharpoons 4NO_2(g)$$

3 The nitrogen(IV) oxide and air are then absorbed into water:

$$4NO_2(g) + O_2(g) + 2H_2O(l) \rightleftharpoons 4HNO_3(aq)$$

The contact process for the manufacture of sulphuric acid

● This takes place in three stages:

1 The combustion of sulphur:

$$S(l) + O_2(g) \rightarrow SO_2(g)$$

2 The reversible oxidation of the sulphur dioxide:

$$SO_2(g) + \tfrac{1}{2}O_2(g) \rightleftharpoons SO_3(g)$$

3 Absorption by the water in 98% sulphuric acid.

● The conditions for stage two, which is exothermic and has fewer gas molecules on the right, are:

Temperature: *425 °C.* A higher temperature would reduce the yield, and a lower one would make the reaction uneconomically slow.

Pressure: *2 atm.* This is enough to force the gases through the plant. A higher pressure is not necessary because the yield is high under these conditions.

Catalyst: *Vanadium(v) oxide.*

Yield: *96% per cycle.* The gases from the catalyst chamber are passed into the absorber containing the 98% sulphuric acid, and all the SO_3 is removed. The gases then go back through another bed of catalyst, giving a final conversion of 99.8%.

Manufacture of aluminium

● Aluminium is too reactive for its oxide to be reduced by carbon monoxide or other cheap reducing agents, so the expensive method of electrolysis of a **molten** ionic compound has to be used.

● The ore contains aluminium oxide (amphoteric) with large impurities of iron oxide (basic) and silicon dioxide (weakly acidic). The ore is treated with a hot 10% solution of sodium hydroxide, which reacts with the amphoteric aluminium oxide to form a solution of sodium aluminate. Iron oxide does not react as it is a base, and silicon dioxide does not

Helpful hints

Aluminium is protected from corroding by its layer of oxide which reforms when it is scratched. It is also less dense than iron. These properties make it ideal as a material for aeroplane wings and fuselage and for drink cans (even though it is more expensive than iron). It is also an excellent conductor of electricity, and so is used in overhead power cables because of these three properties.
As the electrolytic method of manufacture is very expensive, it is economically sensible to recycle aluminium objects.

react because of its giant atomic structure. These solids are filtered off, and carbon dioxide is blown through the solution precipitating aluminium hydroxide. This is obtained by filtration and heated to produce pure aluminium oxide.

- The purified aluminium oxide is dissolved in molten **cryolite**, Na_3AlF_6, at 900 °C.
- The solution is **electrolysed** using carbon anodes dipping into a steel cell lined with carbon, which is the cathode.
- At the **cathode**, aluminium ions are reduced:

$$Al^{3+} + 3e^- \rightarrow Al(l)$$

- The molten aluminium sinks to the bottom of the cell and is siphoned off.
- At the carbon **anode**, oxygen ions are oxidised and react with the anode:

$$2O^{2-} + C(s) \rightarrow CO_2(g) + 4e^-$$

- Because of the expense of this method of manufacture, it is economic to recycle aluminium drink cans.

The production of chlorine

- This is done by the electrolysis of an aqueous solution of sodium chloride (brine).

  At the titanium **anode** (+): NaCl contains Cl^- ions and these are oxidised to Cl_2:

$$2Cl^-(aq) \rightarrow Cl_2(g) + 2e^-$$

  chlorine gas is produced.

- At the steel **cathode** (−): water is ionised in an equilibrium reaction:

$$H_2O(l) \rightleftharpoons H^+(aq) + OH^-(aq)$$

  Na^+ ions are very hard to reduce, and so H^+ ions are preferentially reduced:

$$2H^+(aq) + 2e^- \rightarrow H_2(g).$$

  The removal of the H^+ ions drives the water equilibrium to the right, producing OH^- ions. The overall equation for the reaction at the cathode is:

$$2H_2O + 2e^- \rightarrow 2OH^-(aq) + H_2(g)$$

  Thus sodium hydroxide and hydrogen are produced.

  The anode and cathode compartments are separated by an ion-exchange membrane which allows Na^+ ions to pass through but keeps the chlorine separated from the hydroxide ions that are produced at the cathode.

Helpful hints

Chlorine is used for:
i   water sterilisation
ii the manufacture of organic chlorine compounds, such as the plastic PVC and the herbicide 2,4-D
iii the manufacture of HCl.

Manufacture of sodium chlorate(I)

If the electrolysis of sodium chloride solution is carried out with the solution being stirred, sodium chlorate(I) and hydrogen are obtained. The chlorine produced at the anode disproportionates when in contact with the alkali from the cathode:

$$Cl_2(aq) + 2OH^-(aq) \rightarrow OCl^-(aq) + Cl^-(aq) + H_2O(l)$$

Helpful hints

Sodium chlorate(I) is used:
i   as a domestic bleach
ii as a disinfectant.

Unit 2

Checklist

Before attempting questions on this topic, check that you:

☐ Know the conditions for the manufacture of ammonia.

☐ Know the conditions for the manufacture of nitric acid.

☐ Know the conditions for the manufacture of sulphuric acid.

☐ Can justify the conditions in terms of the economics and the chemistry of these processes.

☐ Know some of the uses of ammonia, nitric acid and sulphuric acid.

☐ Know the conditions and electrode reactions used in the manufacture of aluminium.

☐ Can recall the details of the production of chlorine and sodium chlorate(I) and their uses.

Testing your knowledge and understanding

For the following questions, cover the margin, write your answer, then check to see if you are correct.

● State the conditions used in the manufacture of ammonia.

● State two uses of ammonia.

● State the conditions used for the oxidation of ammonia in the manufacture of nitric acid.

● State the conditions used for the oxidation of sulphur dioxide in the Contact process.

● State two uses of sulphuric acid.

● What property of aluminium oxide is the basis of its purification from bauxite?

● What is the essential condition used in the manufacture of aluminium?

● Write the equations for the reactions at the cathode and at the anode in the manufacture of aluminium.

Answers

A temperature of 350 to 450 °C, a pressure of 200 to 250 atm and an iron catalyst
To make fertilisers, and to make nylon (or nitric acid)
A temperature of 900 °C and a platinum/rhodium catalyst
A temperature of 400 to 450°C, a vanadium(V) oxide catalyst and a pressure of 2 atm
To make fertilisers, and paints (or detergents)
It is amphoteric.
Molten electrolyte
Cathode $Al^{3+} + 3e^- \rightarrow Al$ Anode: $2O^{2-} + C \rightarrow CO_2 + 4e^-$

➪ The answers to the numbered questions are on page 129.

---

1 Explain the economic reasons for the choice of the conditions used in the Haber process.

2 In the Haber process why is the gas mixture cooled after it has left the catalyst chamber?

3 Write the equations for the reactions at the anode and at the cathode in the manufacture of chlorine.

Practice Test: Unit 2

Time allowed 1hr

All questions are taken from parts of previous Edexcel Advanced GCE questions.

The answers are on pages 129–130.

1a Propene, C_3H_6, and but-2-ene, $CH_3CH=CHCH_3$, are in the same homologous series. Explain the term **homologous series**. [3]
  b Draw a representative length of the polymer chain of poly(propene) [2]
  c i Draw the geometric isomers of but-2-ene. [2]
     ii Explain how geometric isomerism arises. [1]
(Total 8 marks)
[June 2001 Unit Test 2 question 2 & May 2002 Unit Test 2 question 5]

2 The rate of any chemical reaction is increased if the temperature is increased.
  a Draw a diagram to represent the Maxwell–Boltzmann distribution of molecular energies at a temperature T_1 and at a higher temperature T_2. [3]
  b Use your diagram and the idea of activation energy to explain why the rate of a chemical reaction increases with increasing temperature. [4]
(Total 7 marks)
[June 2001 Unit Test 2 question 3]

3 Consider the following reaction scheme:

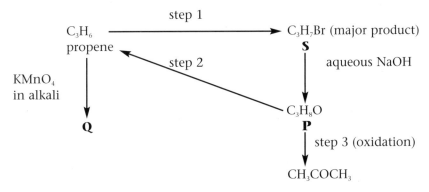

  a i Give the reagent and the conditions needed for step 1. [2]
     ii Give the structural formula of **S**. [1]
  b i Give the structural formula of **P**. [1]
     ii State the **type** of reaction in:
        step 1 [2]
        the conversion of **S** to **P**. [2]
  c i Give the reagent and the conditions needed for step 2. [2]
     ii Give the reagent and the conditions needed for step 3. [3]
  d Give the structural formula of compound **Q**. [1]
(Total 14 marks)
[May 2002 Unit Test 2 question 4]

4a State Hess's Law. [2]
  b Define the term **standard enthalpy change of combustion**. [3]
  c The equation for the combustion of ethanol in air is
$$C_2H_5OH(l) + 3O_2(g) \rightarrow 2CO_2(g) + 3H_2O(l)$$

And the structural representation of this is:

$$H - C - C - O - H + 3O=O \longrightarrow 2O=C=O + 3H-O-H$$

with H, H above the two carbons and H, H below.

i Calculate the enthalpy change for this reaction using the average bond enthalpy values given below.

[3]

Bond	Average bond energy/kJ mol^{-1}	Bond	Average bond energy/kJ mol^{-1}
C—H	+ 412	C—C	+ 348
C—O	+ 360	O—H	+ 463
O=O	+ 496	C=O	+ 743

ii Draw and label an enthalpy level diagram to represent this reaction.

[2]

(Total 10 marks)

[June 2001 Unit Test 2 question 5]

5a The reaction in the Haber Process that is used to produce ammonia is:

$$N_2(g) + 3H_2(g) \rightleftharpoons 2NH_3(g) \qquad \Delta H = -92 \text{ kJ mol}^{-1}$$

    i What temperature is used in the Haber Process? [1]
    ii Justify the use of this temperature. [3]
    iii Name the catalyst used in the Haber process. [1]
    iv How does a catalyst enable a reaction to occur more quickly? [2]

  b Another industrial process is the one which recovers chlorine from HCl, which often is a by-product in organic preparations

$$4HCl(g) + O_2(g) \rightleftharpoons 2Cl_2(g) + 2H_2O(g) \qquad \Delta H = -115 \text{ kJ mol}^{-1}$$

This is a homogeneous dynamic equilibrium reaction.
State the meaning of the terms
    i **homogeneous** [1]
    ii **dynamic equilibrium** [2]

  c State and explain the effect on the position of equilibrium of the reaction in **b** of:
    i decreasing the temperature. [2]
    ii decreasing the volume of the reaction vessel [2]

(Total 14 marks)

[May 2002 question 3 Unit Test 2 & January 2001 question 4 Module Test 2]

6 Aluminium metal is manufactured by a process in which purified bauxite, dissolved in molten cryolite, is electrolysed at 800 °C. Graphite electrodes and a current of about 120 000 amperes are used.
  a Give the ionic equation for the reaction taking place at the anode. [1]
  b Give the ionic equation for the reaction taking place at the cathode. [1]
  c State which of these reactions is an oxidation process. [1]
  d Explain why the anodes need to be replaced frequently. [2]
  e Explain why an electrolyte of pure molten bauxite is not used. [2]

(Total 7 marks)

[June 2001 Unit Test 2 question 7]

[Paper total 60 marks]

3 AS laboratory chemistry

Laboratory chemistry I

● The specification for these tests includes all of Units 1 and 2.

● Unit Test 3B will contain many of the calculations for AS level, and so Topic 1.2 (pages 6 to 12) must be thoroughly revised.

 ### Unit Test 3A: Assessment of practical skills I

☐ This is either internally assessed or a practical exam.

☐ Notes and textbooks are allowed in the tests.

☐ The practical exam will contain some quantitative work, probably a titration or an enthalpy change experiment.

☐ You should know the following gas tests:

● H_2 burns with a squeaky pop

● O_2 relights a glowing spill

● CO_2 turns lime water milky

● NH_3 turns red litmus blue

● Cl_2 bleaches damp litmus; turns KBr solution brown

● NO_2 brown gas which turns starch/iodide paper blue-black

● SO_2 turns acidified potassium dichromate(VI) solution green.

☐ You should know the tests for the following ions:

● CO_3^{2-} Add acid. Test for CO_2.

● HCO_3^- Add to almost boiling water. Test for CO_2.

● SO_4^{2-} Gives white precipitate when dil HCl and $BaCl_2$(aq) are added to the solution.

● HSO_4^- Solution is acidic to litmus, then test as for SO_4^{2-} above.

● SO_3^{2-} Add dil acid to solid and warm. Test for SO_2.

● Halide ion: To solution add dil HNO_3 then $AgNO_3$(aq). White precipitate soluble in dil ammonia indicates chloride; cream precipitate insoluble in dil but soluble in conc ammonia indicates bromide; pale yellow precipitate insoluble in conc ammonia indicates iodide.

● NO_3^- Heat solid. All nitrates give off O_2. All but sodium and potassium nitrates also give off NO_2. Alternatively add aluminium powder and sodium hydroxide solution. Nitrates give off ammonia gas.

● NH_4^+ Add dil NaOH and warm. Test for ammonia.

● Mg^{2+} Gives white precipitate when dil ammonia is added to a solution.

● You must know how to carry out a flame test and the colours obtained.

Ion	Flame colour
Li^+	camine red
Na^+	yellow
K^+	lilac
Ca^{2+}	brick red
Sr^{2+}	crimson red
Ba^{2+}	pale green

 Unit Test 3B: Laboratory chemistry I

❑ This is a written paper, taken by **all** candidates.

❑ It is designed to assess a candidate's ability (related to the topics in Units 1 and 2) to:

   i evaluate information generated from experiments
   ii describe and plan techniques used in the laboratory.

❑ **Tests**

- You should know the tests listed in Unit 3A above.
- You should also know the tests for alkenes, the OH group and the halogen in halogenoalkanes.
- You should be able to deduce the identity of a compound from the results of a series of tests.

❑ **Techniques**

You should be able to describe techniques used in:
- titrations and enthalpy change measurements
- simple organic procedures such as distillation and heating under reflux.

❑ **Planning**

You should be able to:
- Plan a series of tests to determine the identity of an inorganic or organic compound.
- Describe how to make up a solution of known concentration for titrations.
- Plan an experiment to determine the enthalpy of a reaction such as the combustion of a liquid, or the neutralisation of an acid.
- Plan an experiment to follow the progress of a reaction in which there is a change in physical state, such as the production of a gas.

❑ **Calculations**

You should be able to calculate:
- empirical formulae
- reacting masses
- results from titration data
- enthalpy changes from experimental data.

❑ **Evaluation**

You should be able to criticise:
- an experimental plan or apparatus
- the results of an experiment in terms of significant figures, accuracy or experimental error etc.

❑ **Safety**

You should be able to suggest specific safety precautions when a substance is flammable, toxic or irritating.

C=C group	Decolourises brown bromine in hexane
C–OH group	Steamy fumes with PCl_5
Organic halides	Warm with sodium hydroxide solution. Acidify with dilute nitric acid and then test as for ionic halides as on page 55

Practice Test: Unit 3B

Time allowed 1 hr

All questions, except 5, are taken from parts of previous Edexcel Advanced GCE questions.

The answers are on pages 130–132.

1   Complete the table below.

Gas	Reagents or test	Observation expected for a positive result
Hydrogen	Burning splint	
Oxygen	Glowing splint	
Carbon dioxide		
Sulphur dioxide	Potassium dichromate(VI) solution acidified with dilute sulphuric acid.	Solution turns from orange to .....
	Moist blue litmus paper	Turns red and is then bleached white

[6]
(Total 6 marks)
[June 2002 Unit Test 3B question 2]

2 In a series of experiments to investigate the factors that control the rate of a reaction, aqueous hydrochloric acid was added to calcium carbonate in a conical flask placed on an electronic balance. The loss in mass of the flask and its contents was recorded for 15 minutes.

$$CaCO_3(s) + 2HCl(aq) \rightarrow CaCl_2(aq) + H_2O(l) + CO_2(g)$$

Four experiments were carried out.
- Experiments **1**, **3** and **4** were carried out at room temperature (20°C).
- The same mass of calcium carbonate (a large excess) was used in each experiment.
- The pieces of calcium carbonate were the same size in experiments **1**, **2** and **4**.

Experiment	Calcium carbonate	Hydrochloric acid
1	Small pieces	50.0 cm^3 1.00 mol dm^{-3}
2	Small pieces	50.0 cm^3 1.00 mol dm^{-3} heated to 80 °C
3	One large piece	50.0 cm^3 1.00 mol dm^{-3}
4	Small pieces	50.0 cm^3 2.00 mol dm^{-3}

a The results of experiment **1** give the curve shown on the graph below.

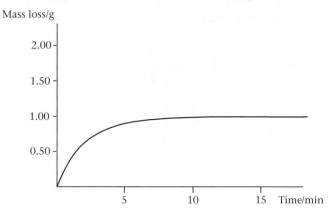

Unit 3

  i   Explain why there is a loss in mass as the reaction proceeds.   [2]
  ii Explain the shape of the curve drawn for experiment **1**.   [2]
b Draw curves on the graph above to represent the results you would expect for experiments **2**, **3** and **4**.   [3]
c i Calculate the mass of calcium carbonate that exactly reacts with 50.0 cm^3 of 1.00 mol dm^{-3} aqueous hydrochloric acid. (Molar mass of CaCO$_3$ = 100 g mol^{-1})   [3]
  ii Based on your answer to **c** part (**i**), suggest a suitable mass of calcium carbonate to use in the experiments. Explain your answer.   [2]

(Total 12 marks)

[January 2002 Unit Test 3B question 2]

3   A student was required to determine the enthalpy change for the reaction between iron and copper sulphate solution. The student produced the following account of their experiment.

> A piece of iron, mass about 3 g, was placed in a glass beaker. Then 50 cm^3 of 0.5 mol dm^{-3} aqueous copper sulphate solution was measured using a measuring cylinder and added to the beaker. The temperature of the mixture was measured immediately before the addition and every minute afterwards until no further change took place
>
> $$Fe + CuSO_4 \rightarrow FeSO_4 + Cu$$
>
> | Timing | Before addition | 1 min | 2 mins | 3 mins | 4 mins | 5 mins |
> |---|---|---|---|---|---|---|
> | Temperature/°C | 16 | 27 | 29 | 26 | 24 | 22 |

a Suggest **two** improvements you would make to this experiment. Give a reason for each of the improvements suggested.   [4]
b In an improved version of **the same experiment** a maximum temperature rise of 15.2 °C occurred when reacting excess iron with 50.0 cm^3 of 0.500 mol dm^{-3} aqueous copper sulphate solution.
  i   Using this data and taking the specific heat capacity of all aqueous solutions as 4.18 J g^{-1} deg^{-1}, calculate the heat change.
  ii Calculate the number of moles of copper sulphate used.   [1]
  iii Calculate the enthalpy change for this reaction in kJ mol^{-1}.   [2]

(Total 8 marks)

[June 2002 Unit Test 3B question 4]

4   A student carried out an experiment to find the percentage of calcium carbonate, CaCO$_3$, in a sample of limestone following his own plan. The student's account of the experiment, results and calculations of the mean titre are given below.

> Account
> I   Mass of piece of limestone = 5.24 g
> II   A pipette was used to transfer 50 cm^3 of 2.00 mol dm^{-3} aqeous hydrochloric acid (an excess) to a 100 cm^3 beaker. The piece of limestone was placed in the beaker and left until there was no more effervescence.
>
> Equation
> $$CaCO_3(s) + 2HCl(aq) \rightarrow CaCl_2(aq) + CO_2(g) + H_2O(l)$$
>
> III The acidic solution in the beaker was filtered into a 250 cm^3 volumetric flask. A small amount of the solid impurity remained in the filter paper. The solution in the volumetric flask was carefully made up to 250 cm^3 with distilled water.
> IV   A pipette was used to tranfer 25.0 cm^3 portions of the acidic solution to conical flasks. The solution was then titrated with 0.100 mol dm^{-3} aqueous sodium hydroxide.
>
> $$HCl(aq) + NaOH(aq) \rightarrow NaCl(aq) + H_2O(l)$$

Results

	1	2	3
Burette reading (final)	14.90	15.40	30.25
Burette reading (at start)	0.00	0.05	15.40
Titre/cm^3	14.90	15.35	14.85

Mean titre = $\dfrac{14.90 + 15.35 + 14.85}{3}$ = 15.033 cm^3

a The accuracy of the student's method was judged to be poor by his teacher. The teacher suggested that the procedure in II could be improved, and that the titres used to calculate the mean were incorrectly chosen.

   i Suggest, with a reason, one improvement to the student's procedure in II. **[2]**

   ii Recalculate a value of the mean making clear which titres you choose. **[2]**

b i Using your answer to **a** part (**ii**), calculate the amount in moles of sodium hydroxide in the mean titre. **[1]**

   ii Hence state the amount in moles of hydrochloric acid in a 25.0 cm^3 portion of the acidic solution transferred in IV. **[1]**

   iii Hence calculate the amount in moles of hydrochloric acid remaining after the reaction in II. **[1]**

   iv Calculate the amount in moles of hydrochloric acid transferred to the beaker in II. **[1]**

   v Hence calculate the amount in moles of hydrochloric acid used in the reaction. **[1]**

   vi Hence calculate the amount in moles of calcium carbonate and the mass of calcium carbonate in the sample of limestone {M_r ($CaCO_3$) = 100} **[2]**

   vii Hence calculate the percentage of calcium carbonate by mass in the sample of limestone. **[1]**

c The burette used in the titrations had an uncertainty for each reading of ±0.05 cm^3.

   i Which of the following should be regarded as the actual value of the titre in titration 3?

     A between 14.80 and 14.90 cm^3:

     B between 14.825 and 14.875 cm^3:

     C between 14.75 and 14.95 cm^3. **[1]**

   ii Suggest one reason why a student may obtain volumes outside the uncertainty of the burette when carrying out a titration. **[1]**

(Total 14 marks)

[June 2001 Unit Test 3B question 4]

5 Propan-2-ol (molar mass 60 g mol-1 and boiling temperature 82 °C) can be prepared by the reaction between 6.15 g of 2-bromopropane (molar mass 123 g mol^{-1} and boiling temperature 59 °C) and excess aqueous 2.0 mol dm^{-3} sodium hydroxide. This reaction is slow at room temperature.

a Describe the procedure, identifying the apparatus that you would use, to prepare a pure sample of propan-2-ol from 6.15 g of 2-bromopropane. **[5]**

b Calculate the minimum volume of sodium hydroxide solution that must be taken to ensure complete reaction of the 2-brompropane. **[2]**

c Your teacher suggested that an 80% yield would be an excellent result. Calculate the mass of pure propan-2-ol that you would need to prepare to obtain this yield. **[2]**

d Suggest one reason why your yield will be below 100%. **[1]**

(Total 10 marks)

Unit 3

4 Periodicity, quantitative equilibria and functional group chemistry

Topic 4.1 Energetics II

Introduction

❏ This topic extends the use of Hess's Law to reactions involving ions.

❏ You must be able to construct a Born Haber cycle.

❏ Lattice energy is crucial to the understanding of ionic bonding, because it is this release of energy that makes the formation of an ionic substance thermodynamically favourable.

❏ Definitions of enthalpy changes must include:
  i the chemical change taking place
  ii the conditions
  iii the amount of substance (reactant or product)
  iv an example of an equation with state symbols, as this may gain marks lost through omission in your word definition.

Things to learn

❏ **The sign** of the value of ΔH tells you the direction of the movement of heat energy:
  ● **positive** (endothermic) for heat flowing into the system
  ● **negative** (exothermic) for heat flowing out of the system.

❏ **Enthalpy of atomisation, ΔH_a,** is the enthalpy change for the production of **one mole of atoms in the gas phase** from the element in its standard state. It is always endothermic.

❏ **Enthalpy of hydration, ΔH_{hyd},** is the enthalpy change when **one mole of gaseous ions** is added to excess water. It is always exothermic.

❏ **Lattice energy, ΔH_{latt},** is the energy change when **one mole of ionic solid** is made from its separate gaseous ions. It is always exothermic.

> ΔH_a for chlorine is the enthalpy change for:
> $\frac{1}{2}Cl_2(g) \rightarrow Cl(g)$
> ΔH_{hyd} for the Na^+ ion is the enthalpy change for:
> $Na^+(g) + aq \rightarrow Na^+(aq)$
> ΔH_{latt} for sodium chloride is the enthalpy change for:
> $Na^+(g) + Cl^-(g) \rightarrow NaCl(s)$

Things to understand

Born–Haber cycle

● This relates the enthalpy of formation of an ionic solid to the enthalpies of atomisation of the elements concerned, the ionisation energies for the formation of the cation, the electron affinity for the formation of the anion and the lattice energy of the ionic substance.

● This can either be drawn as a Hess's Law cycle, or as an enthalpy level diagram (see Figures 4.1 and 4.2).

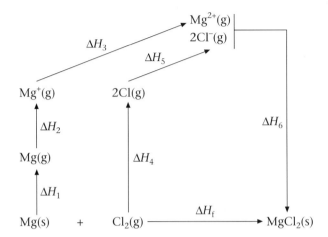

Fig 4.1 *Born–Haber cycle*

Fig 4.2 *Enthalpy level diagram*

ΔH_f of $MgCl_2$ = ΔH_a of Mg (ΔH_1) + 1st ionisation energy of Mg (ΔH_2) + 2nd ionisation energy of Mg (ΔH_3) + 2 × ΔH_a of Cl (ΔH_4) + 2 × electron affinity of Cl (ΔH_5) + ΔH_{latt} of $MgCl_2$ (ΔH_6) = $\Delta H_1 + \Delta H_2 + \Delta H_3 + \Delta H_4 + \Delta H_5 + \Delta H_6$

Any one value can be calculated when all the others are known.

<table>
<tr><td>

Helpful hints

If a bond is partially covalent, the experimental lattice energy (from Born Haber) will be bigger (more exothermic) than the value calculated from a purely ionic model.
The greater the difference between the experimental and theoretical values, the greater the extent of covalency. This will happen if the cation is very polarising (small or highly charged) or the anion is very polarisable (large). See page 13.

</td></tr>
</table>

Lattice energy

● Lattice energy is always exothermic.
● Its value depends upon the strength of the force of attraction (sometimes called the ionic bond) between the ions.
● The strength of this force depends upon the value of the charges and the sum of the radii of the ions:

  i the larger the charge on either or both of the ions, the larger (more exothermic) the lattice energy

  ii the larger the **sum** of the ionic radii, the smaller (less exothermic) the lattice energy.

● Theoretical values of lattice energy can be calculated assuming that the substance is 100% ionic.

Variation of lattice energies in a Group

● **Group 2 sulphates**. As the radius of the cation increases down the Group, the lattice energy decreases. But because the radius of the sulphate ion is much larger than the radii of any of the Group 2 cations, the **sum** of the radii of the cation and the anion alters only **slightly**, and so the lattice energy only decreases **slightly**.

● **Group 2 hydroxides**. As the radius of the cation increases down the Group, the lattice energy decreases. But because the anion is small, and matches the size of the cation (the ionic radii of Ba^{2+} and OH^- are about the same), the **sum** of the radii alters **significantly**, and so the lattice energy decreases by a **large** amount.

Unit 4

Enthalpy of hydration

● Enthalpy of hydration is always exothermic.
● For a cation it is the result of the force of attraction between the ion and the δ^- **oxygen** in the water.
● For an anion it is the result of the force attraction between the ion and the δ^+ **hydrogen** in the water.
● The value of enthalpy of hydration will depend upon the value of the charge on the ion and its radius:
  i   the larger the charge, the larger (more exothermic) the enthalpy of hydration,
  ii the larger the ionic radius, the smaller (less exothermic) the enthalpy of hydration.

Variation of enthalpies of hydration in a Group

The value will become less exothermic as the radius of the cation increases down the Group.

Solubility

The direction of a chemical reaction is partially determined by the value of the enthalpy change. When discussing solubilities, it is assumed that the more exothermic the enthalpy of solution, the more soluble the substance is likely to be. The enthalpy of solution can be estimated from a consideration of lattice energies and enthalpies of hydration of the two ions as can be seen from a Hess's law diagram.

Note that both lattice and hydration enthalpies are negative numbers, so $-\Delta H_{latt}$ is positive

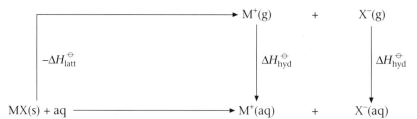

$\Delta H_{soln} = -$ (lattice energy) + (hydration enthalpy of each cation + hydration enthalpy of each anion)

● **Solubility of Group 2 sulphates**
  i   The enthalpy of hydration becomes **much less** as the cations become bigger.
  ii Because the radius of the sulphate ion is much bigger than the radius of any of the cations, the lattice energy only decreases **slightly** – see above.
  iii (i) and (ii) together mean that the enthalpy of solution becomes considerably **less exothermic** and causes the solubility of the sulphates to **decrease** down the Group.
● **Solubility of Group 2 hydroxides**
  i   The enthalpy of hydration becomes **less** as the cations become bigger.
  ii Because the radius of the hydroxide ion is a similar size to that of the cations, the lattice energy **decreases considerably**.
  iii (i) and (ii) together mean that the enthalpy of solution becomes considerably **more exothermic** and causes the solubility of the hydroxides to **increase** down the Group.

Checklist

Before attempting questions on this topic, check that you:

☐ Can define enthalpy of atomisation.

☐ Can define enthalpy of hydration.

☐ Can define lattice energy.

☐ Can construct a Born–Haber cycle.

☐ Can use a Born–Haber cycle to calculate lattice energy or electron affinity.

☐ Understand why the value of ΔH_{latt} from Born–Haber might be more exothermic than the theoretical value.

☐ Can state and explain the change in solubilities of the Group 2 sulphates and hydroxides.

Testing your knowledge and understanding

For the following questions, cover the margin, write your answer, then check to see if you are correct.

Answers
a $MgCO_3$
b $MgCO_3$
c NaF
$Al^{3+} > Mg^{2+} > Ba^{2+} > Na^+$
$Ba(OH)_2$

The answers to the numbered questions are on page 132.

● In each of the following state which compound has the larger (more exothermic) lattice energy:
  a $MgCO_3$ or $BaCO_3$
  b Na_2CO_3 or $MgCO_3$
  c NaF or NaCl.

● Arrange the following ions in order of decreasing (less exothermic) enthalpy of hydration: Na^+, Mg^{2+}, Ba^{2+} and Al^{3+}

● State the formula of the most soluble Group 2 hydroxide.

1 The following data, in $kJ\ mol^{-1}$, should be used in this question.

Enthalpies of atomisation: calcium +193; chlorine +121.

Ionisation energies for calcium: 1st +590; 2nd +1150

Electron affinity for Cl(g): –364

Lattice energies: for $CaCl_2(s)$ –2237; for CaCl(s) –650 (estimated).

a Construct a Born–Haber cycle for:

  i the formation of CaCl(s)
  ii the formation of $CaCl_2(s)$.

b Calculate the standard enthalpy of formation of:

  i $CaCl_2(s)$
  ii CaCl(s)
  iii hence calculate the enthalpy of the reaction:
  $$2CaCl(s) \rightarrow Ca(s) + CaCl_2(s)$$
  and comment on the fact that CaCl(s) does not exist.

2 Comment on the values of the lattice energies, in $kJ\ mol^{-1}$, given below:

	Experimental	Theoretical
NaF	–918	–912
MgI_2	–2327	–1944

Unit 4

Topic 4.2 The Periodic Table II

 ## Introduction

☐ You will need to look again at Topic 1.4.

☐ The elements become **more metallic** down a Group. This means that their compounds, such as their chlorides, become more ionic and their oxides become more basic.

☐ The elements become **less metallic** from left to right across a Period. This means that their chlorides become more covalent and their oxides become more acidic.

☐ There are many formulae and equations that must be learnt in this Topic.

☐ The properties of elements and their compounds change steadily from the top to the bottom of a Group and from left to right across a Period.

 ## Things to learn and understand

Variation of properties across Period 3

A table is given below showing the formulae of the products of the reactions of the elements with oxygen, chlorine and water:

	Na	Mg	Al	Si	P	S	Cl	Ar
Oxygen	Na_2O or Na_2O_2 [1]	MgO	Al_2O_3	SiO_2	P_4O_{10}	SO_2	No reaction	No reaction
Chlorine	NaCl	$MgCl_2$	$AlCl_3$ [2]	$SiCl_4$	PCl_3 PCl_5 [3]	S_2Cl_2		No reaction
Water	NaOH + H_2	MgO + H_2 [4]	No reaction	No reaction	No reaction	No reaction	HCl + HOCl	No reaction

Notes: [1] with excess oxygen sodium peroxide is formed, [2] aluminium chloride is covalent, [3] with excess chlorine, [4] with steam.

☐ **Metal hydroxides and oxides**
- NaOH   alkaline: $NaOH(s) + aq \rightarrow Na^+(aq) + OH^-(aq)$
- $Mg(OH)_2$ basic: $Mg(OH)_2(s) + 2H^+(aq) \rightarrow Mg^{2+}(aq) + 2H_2O(l)$
- MgO   basic: $MgO(s) + 2H^+(aq) \rightarrow Mg^{2+}(aq) + H_2O(l)$
- $Al(OH)_3$ amphoteric: $Al(OH)_3(s) + 3H^+(aq) \rightarrow Al^{3+}(aq) + 3H_2O(l)$
  $Al(OH)_3(s) + 3OH^-(aq) \rightarrow Al(OH)_6^{3-}(aq)$
- Al_2O_3 amphoteric: $Al_2O_3(s) + 6H^+(aq) \rightarrow 2Al^{3+}(aq) + 3H_2O(l)$
  $Al_2O_3(s) + 6OH^-(aq) + 3H_2O(l) \rightarrow 2Al(OH)_6^{3-}(aq)$

☐ **Non-metal oxides**
- SiO_2 weakly acidic: $SiO_2(s) + 2NaOH(l) \rightarrow Na_2SiO_3(l) + H_2O(g)$
- P_4O_6 weakly acidic: $P_4O_6 + 6H_2O \rightarrow 4H_3PO_3 \rightleftharpoons 4H^+(aq) + 4H_2PO_3^-$
- P_4O_{10} strongly acidic: $P_4O_{10} + 6H_2O \rightarrow 4H_3PO_4 \rightarrow 4H^+(aq) + 4H_2PO_4^-$

- SO_2 weakly acidic: $SO_2 + H_2O \rightarrow H_2SO_3 \rightleftharpoons H^+(aq) + HSO_3^-$
- SO_3 strongly acidic: $SO_3 + H_2O \rightarrow H_2SO_4 \rightarrow H^+(aq) + HSO_4^-$

☐ Chlorides

- NaCl ionic solid, dissolves in water:
$$NaCl(s) + aq \rightarrow Na^+(aq) + Cl^-(aq)$$
- $MgCl_2$ ionic solid, dissolves in water:
$$MgCl_2(s) + aq \rightarrow Mg^{2+}(aq) + 2Cl^-(aq)$$
- Hydrated $AlCl_3$ ionic solid, deprotonated by water:
$$[Al(H_2O)_6]^{3+}(aq) + H_2O \rightarrow [Al(H_2O)_5(OH)]^{2+}(aq) + H_3O^+(aq)$$
- Anhydrous $AlCl_3$ covalent solid, reacts with water:
$$AlCl_3(s) + 3H_2O(l) \rightarrow Al(OH)_3(s) + 3HCl(aq)$$
- $SiCl_4$ covalent liquid, reacts with water:
$$SiCl_4(l) + 2H_2O(l) \rightarrow SiO_2(s) + 4HCl(aq)$$
- PCl_3 covalent liquid, reacts with water:
$$PCl_3(l) + 3H_2O(l) \rightarrow H_3PO_3(aq) + 3HCl(aq)$$
- PCl_5 covalent solid, reacts with water:
$$PCl_5(s) + 4H_2O(l) \rightarrow H_3PO_4(aq) + 5HCl(aq)$$

Variation of properties down Group 4

☐ The **metallic character** increases down the Group:
carbon and silicon are non-metallic (not malleable; only graphite conducts electricity)
germanium is semi-metallic (semi-conductor).
tin and lead are metallic (malleable and electrical conductors).

☐ **The +2 oxidation state becomes more stable relative to the +4 state**. Lead(IV) is oxidising and is itself reduced to lead(II), whereas tin(II) is reducing and is itself oxidised to tin(IV). This is illustrated by:
- lead(IV) oxide oxidises concentrated hydrochloric acid **to** chlorine:
$$PbO_2 + 4HCl \rightarrow PbCl_2 + Cl_2 + H_2O$$
- tin(II) ions are oxidised to tin(IV) ions **by** chlorine:
$$Sn^{2+} + Cl_2 \rightarrow Sn^{4+} + 2Cl^-$$

☐ **Tetrachlorides**. Both CCl_4 and $SiCl_4$ are covalent and are tetrahedral molecules owing to the repulsion of the four bond pairs of electrons. $SiCl_4$ is rapidly hydrolysed by water. A lone pair of electrons from the oxygen in the water forms a dative bond into an empty 3d orbital in the silicon atom. The energy released is enough to overcome the activation energy barrier involved in the breaking of the Si–Cl bond. Carbon's bonding electrons are in the 2nd shell and there are no 2d orbitals. The empty orbitals in the 3rd shell are of too high an energy to be used in bonding, and equally important is that carbon is such a small atom that water molecules are prevented from reaching it by the four much larger chlorine atoms arranged tetrahedrally around it.

☐ **Acidity of the oxides**. This decreases down the Group:

- Carbon dioxide is weakly acidic, and reacts with metal oxides and alkalis:
$$CO_2(g) + 2OH^-(aq) \rightarrow CO_3^{2-}(aq) + H_2O(l)$$
- Silicon dioxide is very weakly acidic, and reacts with molten sodium hydroxide:
$$SiO_2(s) + 2NaOH(l) \rightarrow Na_2SiO_3(l) + H_2O(g)$$
- Lead oxides (and hydroxide) are amphoteric, and react with aqueous acids and alkalis, but lead(II) oxide is the strongest base of the Group 4 oxides:
$$PbO(s) + 2H^+(aq) \rightarrow Pb^{2+}(aq) + H_2O(l)$$
$$PbO(s) + 2OH^-(aq) + H_2O \rightarrow Pb(OH)_4^{2-}(aq)$$

Helpful hints

Ionic chlorides dissolve in water to form hydrated ions.
These hydrated ions may deprotonate, if the surface charge density of the metal ion is high.
Covalent chlorides (of Period 3) react with water to form HCl and either the oxide or the oxoacid.

Helpful hints

The trend in metallic character is caused by the decrease in ionisation energy as the radius of the atom increases, making it energetically favourable for the formation of ionic bonds.

Unit 4

Checklist

Before attempting the questions on this topic, check that, for the Period 3 elements, you:

☐ Can write equations for the reactions of the elements with oxygen and with chlorine.

☐ Can write equations for the reaction of the elements with water.

☐ Can state and explain the acid/base nature of the metal hydroxides.

☐ Can state and explain the acid/base nature of the non-metal oxides.

☐ Know the formula of their chlorides and their reactions with water, and can relate the reactions to their bonding.

Before attempting the questions on this topic, check that, for the Group 4 elements, you:

☐ Know why the elements become more metallic down the Group.

☐ Know that the +2 oxidation state becomes more stable down the Group.

☐ Know why $SiCl_4$ is rapidly hydrolysed by water, whereas CCl_4 is unreactive with water.

☐ Can write equations to show the acid/base character of the oxides of carbon, silicon and lead.

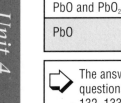 *Testing your knowledge and understanding*

Answers
PbO and PbO_2
PbO

⇨ The answers to the numbered questions are on pages 132–133.

For the following questions, cover the margin, write your answer, then check to see if you are correct.

● State the formulae of lead(II) oxide and lead(IV) oxide.

● State the formula of the most basic Group 4 oxide.

1 a Write equations for the changes caused by the addition of water to:
   i sodium chloride
   ii silicon tetrachloride
   iii phosphorus pentachloride.
  b Relate these reactions to the bonding in the chlorides.
  c Explain why silicon tetrachloride, $SiCl_4$, reacts rapidly with water, but tetrachloromethane, CCl_4, does not react even when heated with water to $100\,^{\circ}C$.

2 Write two ionic equations to show that aluminium hydroxide is amphoteric.

3 Write ionic equations for the reactions, if any, of:
  a aqueous acid with:
   i magnesium hydroxide
   ii lead monoxide
   iii sulphur dioxide
  b aqueous alkali with:
   i magnesium hydroxide
   ii lead monoxide
   iii sulphur dioxide.

4 a Write an equation to show lead(IV) oxide acting as oxidising agent.
  b Write an equation to show aqueous tin(II) ions acting as a reducing agent.

Unit 4

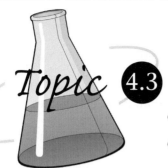

Topic **4.3** Chemical equilibria II

 Introduction

❑ Concentration, in the context of equilibrium, is always measured in mol dm^{-3}, and the concentration of a substance A is written as [A].

❑ The common errors in this topic are to use moles not concentrations and initial values not equilibrium values when substituted in expressions for K_c.

❑ Calculations of K must be set out clearly, showing each step.

❑ A homogeneous equilibrium is one in which all the substances are in the same phase, whereas in a heterogeneous equilibrium there are two or more phases.

Things to learn and understand

The equilibrium constant K_c

The equilibrium constant, measured in terms of concentrations, is found from the chemical equation.

● For a reaction:

$$m\text{A} + n\text{B} \rightleftharpoons x\text{C} + y\text{D}$$

where m, n, x and y are the stoichiometric numbers in the equation:
$$K_c = \frac{[\text{C}]^x_{eq} \, [\text{D}]^y_{eq}}{[\text{A}]^m_{eq} \, [\text{B}]^n_{eq}}$$

All the concentrations are equilibrium values.

● In a heterogeneous equilibrium a solid substance does not appear in the expression for K_c.

● K_c is only equal to the quotient $\dfrac{[\text{C}]^x \, [\text{D}]^y}{[\text{A}]^m \, [\text{B}]^n}$

when the system is at equilibrium.

● If the quotient does **not** equal K_c, the system is **not** in equilibrium and will **react** until it reaches equilibrium when there will be no further change in any of the concentrations.

● K_c has units. For the reaction:
$$\text{N}_2 + 3\text{H}_2 \rightleftharpoons 2\text{NH}_3$$
$$K_c = \frac{[\text{NH}_3]^2_{eq}}{[\text{N}_2]_{eq} \, [\text{H}_2]^3_{eq}}$$

the units of K_c are $\dfrac{(\text{mol dm}^{-3})^2}{(\text{mol dm}^{-3}) \, (\text{mol dm}^{-3})^3} = \dfrac{1}{(\text{mol dm}^{-3})^2} = \text{mol}^{-2} \, \text{dm}^6$

> The value of K_c is specific to a reaction – and to the equation used to represent that reaction – and can only be altered by changing the temperature.

Unit 4

Calculation of K_c

This must be done using equilibrium concentration values, not initial concentration values. The calculation is done in 5 steps:

i Draw up a table and fill in the initial number of moles, the change for each substance and the equilibrium number of moles of each substance.

ii Convert equilibrium moles to concentration in mol dm^{-3}.

iii State the expression for K_c.

iv Substitute equilibrium concentrations into the expression for K_c.

v Work out the units for K_c and add them to your answer.

Worked example

An important reaction in the blast furnace is the formation of carbon monoxide from carbon and carbon dioxide. This reaction is an example of a heterogeneous equilibrium.

When 1.0 mol of carbon dioxide was heated with excess carbon to a temperature of 700 °C in a vessel of volume 20 dm^3, 95% of the carbon dioxide reacted fo form carbon monoxide. Calculate K_c for

$$C(s) + CO_2(g) \rightleftharpoons 2CO(g)$$

Answer:

	$CO_2(g)$	$CO(g)$	Units
Initial amount	1.0	0	mol
Change	−0.95	+0.95 × 2	mol
Equilibrium amount	1.0 − 0.95 = 0.05	1.9	mol
Equilibrium concentration	0.05/20 = 0.0025	1.9/20 = 0.095	mol dm^{-3}

$$K_c = \frac{[CO]^2_{eq}}{[CO_2]_{eq}} = \frac{(0.095)^2}{0.0025} \frac{(\text{mol dm}^{-3})^2}{\text{mol dm}^{-3}} = 3.6 \text{ mol dm}^{-3}$$

- C(s) does not appear in the K_c expression because it is a solid.
- The amount of CO made is twice the amount of CO_2 reacted.

Partial pressure

- The partial pressure of a gas A, p_A, in a mixture is the pressure that the gas would exert if it alone filled the container. It is calculated from the expression:

  partial pressure of a gas = mole fraction of that gas × total pressure

  where the mole fraction = $\dfrac{\text{the number of moles of that gas}}{\text{the total number of moles of gas}}$

- The total pressure, P, is equal to the sum of the partial pressures of each gas in the mixture.

The equilibrium constant K_p

The equilibrium constant, measured in terms of partial pressure, only applies to reactions involving gases. **Solids or liquids** do not appear in the expression for K_p.

For a reaction:

$$mA(g) \rightleftharpoons xB(g) + yC(g)$$

where m, x and y are the stoichiometric numbers in the equation:

$$K_p = \frac{(p_B)^x_{eq} \times (p_C)^y_{eq}}{(p_A)^m_{eq}}$$

> As carbon is a solid it is ignored in the expression.

All partial pressures are equilibrium values.

So for the reaction:

$$CO_2(g) + C(s) \rightleftharpoons 2CO(g)$$

$$K_p = \frac{p(CO)^2_{eq}}{p(CO_2)_{eq}} \text{ and its units are } \frac{atm^2}{atm} = atm$$

Calculations involving K_p

- The calculation of K_p from equilibrium data is done in a similar way to those involving K_c:

    i Draw up a table and fill in the initial number of moles, the change for each substance, the equilibrium number of moles of each substance and the total number of moles at equilibrium.

    ii Convert equilibrium moles to mole fraction.

    iii Multiply the mole fraction of each by the **total** pressure.

    iv State the expression for K_p.

    v Substitute equilibrium concentrations into the expression for K_p.

    vi Work out the units for K_p and add them to your answer.

Worked example

0.080 mol of PCl_5 was placed in a vessel and heated to 175 °C. When equilibrium had been reached, it was found that the total pressure was 2.0 atm and that 40% of the PCl_5 had dissociated. Calculate K_p for the reaction:

$$PCl_5(g) \rightleftharpoons PCl_3(g) + Cl_2(g)$$

Answer: As 40% had dissociated, 60% was left. Thus at equilibrium:
$0.60 \times 0.080 = 0.048$ mol of PCl_5 was present and therefore
$0.080 - 0.048 = 0.032$ mol of PCl_5 had reacted.

	PCl_5	PCl_3	Cl_2	Total
Initial amount/mol	0.080	0	0	
Change/mol	−0.032	+0.032	+0.032	
Equilibrium amount /mol	0.60×0.080 = 0.048	0.032	0.032	0.112
Mol fraction	0.048/0.112 = 0.429	0.032/0.112 = 0.286	0.032/0.112 = 0.286	
Partial pressure /atm	0.429×2 = 0.857	0.286×2 = 0.571	0.286×2 = 0.571	

$$K_c = \frac{p(PCl_3)_{eq} \, p(Cl_2)_{eq}}{p(PCl_5)_{eq}} = \frac{0.571 \text{ atm} \times 0.571 \text{ atm}}{0.857 \text{ atm}} = 0.38 \text{ atm}$$

Helpful hints

K only equals the quotient when the system is at equilibrium. If the quotient is greater than K, the reaction will move to the left until the two are equal. If the quotient is less than K, the reaction will move to the right until the two are equal. Temperature changes alter the value of K (unless $\Delta H = 0$). Concentration and pressure changes alter the value of the quotient.

Variation of K with conditions

- **Temperature**. This is the only factor that alters the value of K.

    i If a reaction is exothermic left to right, an increase in temperature will lower the value of K. This means that the position of equilibrium will shift to the left (the endothermic direction).

    ii If a reaction is endothermic left to right, an increase in temperature will increase the value of K. This means that the position of equilibrium will shift to the right.

- **Catalyst**. This neither alters the value of K nor the position of equilibrium. It speeds up the forward and the reverse reactions equally. Thus it causes equilibrium to be reached more quickly.

Unit 4

● **Concentration**. A change in the concentration of one of the substances in the equilibrium mixture will **not** alter the value of K, but it will alter the value of the quotient. Therefore the reaction will no longer be at equilibrium. It will react until the value of the quotient once again equals K. If the concentration of a reactant on the left hand side of the equation is increased, the position of equilibrium will move to the right.

● **Pressure**. A change in pressure does not alter K. If there are more gas molecules on one side than the other, the value of the quotient will be altered by a change in pressure. Therefore the reaction will no longer be at equilibrium, and will react until the value of the quotient once again equals K. If there are more gas molecules on the right of the equation and the pressure is increased, the position of equilibrium will move to the left. For example, if the pressure is increased on the equilibrium $N_2O_4(g) \rightleftharpoons 2NO_2(g)$, the position of equilibrium will shift to the left as there are fewer gas molecules on the left side of the equation.

 Checklist

Before attempting questions on this topic, check that you:

☐ Can define the partial pressure of a gas.

☐ Can deduce the expression for K_c and its units given the equation.

☐ Can deduce the expression for K_p and its units given the equation.

☐ Can calculate the value of K_c given suitable data.

☐ Can calculate the value of K_p given suitable data.

☐ Know not to include values for solids and liquids in the expression for K_p.

☐ Know that **only** temperature can alter the value of K, and how it will affect the position of equilibrium.

 Testing your knowledge and understanding

For the following questions, cover the margin, write your answer, then check to see if you are correct.

● Dry air at a pressure of 104 kN m^{-2} contains 78.1% nitrogen 21.0% oxygen and 0.9% argon by moles. Calculate the partial pressures of each gas.

● Write the expression for K_c, stating its units, for the following reactions:
  a $2SO_2(g) + O_2(g) \rightleftharpoons 2SO_3(g)$

  b $2HCl(g) + \frac{1}{2}O_2(g) \rightleftharpoons H_2O(g) + Cl_2(g)$

● Write the expression for K_p, stating its units, for the following reactions:
  a $CO(g) + 2H_2(g) \rightleftharpoons CH_3OH(g)$

  b $Fe_2O_3(s) + CO(g) \rightleftharpoons 2FeO(s) + CO_2(g)$

1 When a 0.0200 mol sample of sulphur trioxide was introduced into a vessel of volume 1.52 dm^3 at 1000 °C, 0.0142 mol of sulphur trioxide was found to be present after equilibrium had been reached.

Answers

$p(N_2) = 0.781 \times 104 = 81.2$ kN m^{-2}
$p(O_2) = 0.210 \times 104 = 21.8$ kN m^{-2}
$p(Ar) = 0.009 \times 104 = 0.9$ kN m^{-2}

a $K_c = \dfrac{[SO_3]^2}{[SO_2]^2[O_2]} = $ mol^{-1} dm^3

b $K_c = \dfrac{[H_2O][Cl_2]}{[HCl]^2[O_2]^{1/2}} = $ mol$^{-1/2}$ dm$^{3/2}$

a $K_p = \dfrac{p(CH_3OH)}{p(CO) \times p(H_2))^2}$ atm^{-2}

b $K_p = \dfrac{p(CO_2)}{p(CO)}$ no units

The answers to the numbered questions are on page 133.

Calculate the value of K_c for the reaction:
$$2SO_3(g) \rightleftharpoons 2SO_2(g) + O_2(g)$$

2 1.0 mol of nitrogen(II) oxide, NO, and 1.0 mol of oxygen were mixed in a container and heated to 450 °C. At equilibrium the number of moles of oxygen was found to be 0.70 mol. The total pressure in the vessel was 4.0 atm. Calculate the value of K_p for the reaction:
$$2NO(g) + O_2(g) \rightleftharpoons 2NO_2(g)$$

3 This question concerns the equilibrium reaction:
$$2SO_2(g) + O_2(g) \rightleftharpoons 2SO_3(g) \quad \Delta H = -196 \text{ kJ mol}^{-1}$$
$$K_c = 3 \times 10^4 \text{ mol}^{-1} \text{ dm}^3 \text{ at } 450 \text{ °C}$$

a 2 mol of sulphur dioxide, 1 mol of oxygen and 2 mol of sulphur trioxide were mixed in a vessel of volume 10 dm^3 at 450 °C in the presence of a catalyst. State whether these substances are initially in equilibrium. If not, explain which way the system would react.

b State and explain the effect on the value of K_c and on the position of equilibrium of:

   i decreasing the temperature

   ii decreasing the pressure

   iii adding more catalyst.

$\mathcal{T}opic$ Acid–base equilibria

 ## *Introduction*

- It is essential that you can write the expression for K_a of a weak acid.
- Make sure that you know how to use your calculator to evaluate logarithms, and how to turn pH and pK_a values into [H$^+$] and K_a values.
- This means using the lg key for $\log_{10}$ and 10^x key for inverse $\log_{10}$.
- Give pH values to 2 decimal places.
- Buffer solutions do **not** have a **constant** pH. They resist changes in pH.
- You may use H$^+$, H$^+$(aq) or H$_3$O$^+$ as the formula of the hydrogen ion.

Things to learn

☐ A **Brønsted–Lowry acid** is a substance that donates an H$^+$ ion (a proton) to another species.

☐ A **monobasic** (monoprotic) acid contains, per molecule, one hydrogen atom which can be donated as an H$^+$ ion. A dibasic acid contains two.

☐ A **Brønsted–Lowry base** is a substance that accepts an H$^+$ ion from another species.

☐ $K_w = [\text{H}^+(\text{aq})][\text{OH}^-(\text{aq})] = 1.0 \times 10^{-14} \text{ mol}^2 \text{ dm}^{-6}$ at 25 °C.

☐ A **neutral** solution is one where $[\text{H}^+(\text{aq})] = [\text{OH}^-(\text{aq})]$.

☐ **pH** $= -\log_{10}[\text{H}^+(\text{aq})]$, or more accurately $= -\log_{10}([\text{H}_3\text{O}^+(\text{aq})]/\text{mol dm}^{-3})$.

❏ **pOH** = $-\log_{10}[OH^-(aq)]$.

❏ **pH + pOH** = 14 at 25 °C.

❏ A **strong** acid is totally ionised in aqueous solution, and a **weak** acid is only partially ionised in aqueous solution.

❏ K_a for a weak acid, HA, = $[H^+] [A^-] / [HA]$.

❏ **pK_a** = $-\log_{10} K_a$.

❏ A **buffer** solution is a solution of known pH which has the ability to resist changing pH when small amounts of acid or base are added.

 Things to understand

Conjugate acid/base pairs

❏ These are linked by an H^+ ion:

Acid $- H^+ \rightarrow$ its conjugate base:
$$CH_3COOH \text{ (acid) } - H^+ \rightarrow CH_3COO^- \text{ (conjugate base)}$$
Base $+ H^+ \rightarrow$ its conjugate acid:
$$NH_3 \text{ (base) } + H^+ \rightarrow NH_4^+ \text{ (conjugate acid)}$$

❏ When sulphuric acid is added to water, it acts as an acid and the water acts as a base:

$$H_2SO_4 + H_2O \rightarrow \qquad H_3O^+ \qquad + \qquad HSO_4^-$$
acid   base   conjugate acid of H_2O conjugate base of H_2SO_4

❏ Hydrogen chloride gas acts as an acid when added to ammonia gas:

$$HCl + NH_3 \rightarrow \qquad NH_4^+ \qquad + \qquad Cl^-$$
acid   base   conjugate acid of NH_3 conjugate base of HCl

pH scale

❏ Water partially ionises

$$H_2O \rightleftharpoons H^+(aq) + OH^-(aq)$$

As the $[H_2O]$ is very large and so is effectively constant, its value can be incorporated into the K for the reaction:
$$K_w = [H^+] [OH^-] = 1.0 \times 10^{-14} \text{ mol}^2 \text{ dm}^{-6} \text{ at 25 °C}$$
or   pH + pOH = 14 at 25 °C

❏ So at **25 °C**:

● A neutral solution is one where $[H^+] = [OH^-]$, and the pH = 7.
● An acidic solution is one where $[H^+] > [OH^-]$, and the pH < 7.
● An alkaline solution is where $[H^+] < [OH^-]$, and the pH > 7.

❏ **pH of strong acids and bases**

● For a strong monobasic acid:
$$pH = -\log_{10}[acid]$$
Therefore the pH of a 0.321 mol dm^{-3} solution of HCl = $-\log_{10} (0.321)$ = 0.49
● For a strong alkali with one OH^- ion in the formula:
$$pOH = -\log_{10}[alkali] \text{ and}$$
$$pH = 14 - pOH$$

Helpful hints

If the acidity of a solution increases, $[H^+]$ increases but the pH decreases.

Helpful hints

The 2nd ionisation of a dibasic acid such as H_2SO_4 is weak. So $[H^+]$ for 0.321 mol dm^{-3} H_2SO_4 is **not** 0.642 mol dm^{-3}, but is only 0.356 mol dm^{-3}.

Helpful hints

For an alkali with 2 OH per formula, the $[OH^-]$ is 2 x the [alkali].
The pH of a 0.109 mol dm^{-3} solution of Ba(OH)$_2$ is
pOH = $-\log_{10}(2 \times 0.109)$ = 0.66
Therefore pH = 14 − 0.66 = 13.34
Remember that acids have a pH < 7 and alkalis a pH > 7.

Therefore the pH of a 0.109 mol dm^{-3} solution of NaOH is calculated:

$$pOH = -\log_{10}(0.109) = 0.96$$

Therefore pH = 14 − 0.96 = 13.04

pH of weak acids

● Weak acids are partially ionised. A general equation for this is:

$$HA(aq) \rightleftharpoons H^+(aq) + A^-(aq)$$

● The acid produces **equal** amounts of H$^+$ and A$^-$, so that:

$$[H^+(aq)] = [A^-(aq)]$$

● This enables either the pH of a weak acid solution or the value of K_a of a weak acid to be calculated.

Worked example

Calculate the pH of a 0.123 mol dm^{-3} solution of an acid HA which has $K_a = 4.56 \times 10^{-5}$ mol dm^{-3}:

$$HA(aq) \rightleftharpoons H^+(aq) + A^-(aq)$$

Answer: $\quad K_a = \dfrac{[H^+(aq)] \; [A^-(aq)]}{[HA(aq)]} = \dfrac{[H^+(aq)]^2}{[HA(aq)]}$

$[H^+(aq)] = \sqrt{(K_a \cdot [HA(aq)])}$

$\qquad\qquad = \sqrt{(4.56 \times 10^{-5} \times 0.123)} = 2.37 \times 10^{-3}$ mol dm^{-3}

$pH = -\log_{10}(2.37 \times 10^{-3}) = 2.63$

Titration curves

● These are drawn either for the addition of base to 25 cm^3 of a 0.1 mol dm^{-3} solution of acid until the base is in excess, or for addition of acid to a base (see Figure 4.3).

● If the acid and base are of the same concentration, the end point is at 25 cm^3.

● There are several pH values to remember.

Helpful hints

At **half way** to the end point (12.5 cm^3 if the end point is at 25 cm^3), the curve for a weak acid and a strong base will be almost horizontal at a pH of about 5. This is when the solution is buffered and [HA] = [A$^-$]. Thus at the half way pH, $pK_a = pH$ for the weak acid.

	Starting pH	End point pH	Vertical pH range	Final pH
Strong acid/strong base	1	7	3.5 to 10.5	13
Weak acid/strong base	3	9	7 to 10.5	13
Strong acid/weak base	1	5	3.5 to 7	11

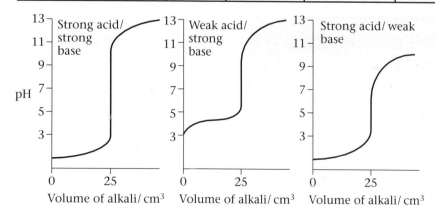

Fig 4.3 Titration curves

Unit 4

Indicators

- For use in a titration, an indicator must change colour within the pH of the **vertical** part of the titration curve.

	Vertical range of pH	Suitable indicator
Strong acid and base	3.5 to 10.5	Methyl orange or phenolphthalein
Weak acid/strong base	7 to 10.5	Phenolphthalein
Strong acid/weak base	3.5 to 7	Methyl orange

Buffer solutions

- These consist of an acid/base conjugate pair, e.g. a weak acid and its salt such as CH_3COOH/CH_3COONa, or a weak base and its salt, e.g. NH_3/NH_4Cl.

- To be able to resist pH changes, the concentration of **both** the acid and its conjugate base must be **similar**.

- Consider a buffer of ethanoic acid and sodium ethanoate. The salt is **fully** ionised:

$$CH_3COONa(aq) \rightarrow CH_3COO^-(aq) + Na^+(aq)$$

The weak acid is **partially** ionised:

$$CH_3COOH(aq) \rightleftharpoons H^+(aq) + CH_3COO^-(aq)$$

The CH_3COO^- ions from the salt suppress most of the ionisation of the acid, and so **both** $[CH_3COOH(aq)]$ and $[CH_3COO^-(aq)]$ are large.

If H^+ ions are added to the solution, almost all of them are removed by reaction with the **large reservoir** of CH_3COO^- ions from the salt:

$$H^+(aq) + CH_3COO^-(aq) \rightarrow CH_3COOH(aq)$$

If OH^- ions are added, almost all of them are removed by reaction with the **large reservoir** of CH_3COOH molecules of the weak acid:

$$OH^-(aq) + CH_3COOH \rightarrow CH_3COO^-(aq) + H_2O(l)$$

Calculation of the pH of a buffer solution

Consider a buffer made from a weak acid, HA, and its salt NaA.
The acid is partially ionised:

$$HA(aq) \rightleftharpoons H^+(aq) + A^-(aq)$$

$$\text{Thus } K_a = \frac{[H^+(aq)]\ [A^-(aq)]}{[HA(aq)]}$$

The salt is totally ionised, and suppresses the ionisation of the acid,

$$\text{Therefore } [A^-(aq)] = [\text{salt}]$$

$$\text{and } [HA(aq)] = [\text{weak acid}]$$

$$\text{Thus } K_a = \frac{[H^+(aq)]\ [\text{salt}]}{[\text{weak acid}]}$$

$$\text{or } [H^+(aq)] = \frac{K_a\ [\text{weak acid}]}{[\text{salt}]}$$

$$pH = -\log[H^+(aq)]$$

Helpful hints

If the concentrations of the weak acid and its salt are the same, then:
$[H^+(aq)] = K_a$, and
$pH = pK_a$.

Worked example

Calculate the pH of 500 cm³ of a solution containing 0.121 mol of ethanoic acid and 0.100 mol of sodium ethanoate.

pK_a for ethanoic acid $= 4.76$

Answer: $pK_a = 4.76$,

Therefore $K_a = $ inverse $\log(-pK_a) = 1.74 \times 10^{-5}$ mol dm⁻³

[weak acid] $= 0.121 \div 0.500 = 0.242$ mol dm⁻³

[salt] $\qquad = 0.100 \div 0.500 = 0.200$ mol dm⁻³.

$[H^+(aq)] = K_a \cdot \dfrac{[\text{weak acid}]}{[\text{salt}]} = \dfrac{1.74 \times 10^{-5} \times 0.242}{0.200} = 2.10 \times 10^{-5}$ mol dm⁻³

$pH = -\log_{10}[H^+(aq)] = 4.68$

Enthalpy of neutralisation

When a strong acid is neutralised by a solution of a strong base, $\Delta H^0_{neut} = -57$ kJ mol⁻¹, because the reaction for strong acids with strong bases is:

$$H^+(aq) + OH^-(aq) = H_2O(l)$$

The value for the neutralisation of a weak acid is less because energy has to be used to ionise the molecule to form H⁺ ions.

 Checklist

Before attempting the questions on this topic, check that you:

❑ Can identify acid/base conjugate pairs.

❑ Can define pH and K_w.

❑ Can define K_a and pK_a for weak acids.

❑ Understand what is meant by the terms strong and weak as applied to acids and bases.

❑ Can calculate the pH of solutions of strong acids, strong bases and weak acids.

❑ Can recall the titration curves for the neutralisation of strong and weak acids.

❑ Can use the curve to calculate the value of K_a for a weak acid.

❑ Understand the reasons for the choice of indicator in an acid/base titrations.

❑ Can define a buffer solution, explain its mode of action and calculate its pH.

 Testing your knowledge and understanding

For the following questions, cover the margin, write your answer, then check to see if you are correct.

● Identify the acid/base conjugate pairs in the reaction:

$$H_2SO_4 + CH_3COOH \rightleftharpoons CH_3COOH_2^+ + HSO_4^-$$

● Calculate the pH of:

a 0.11 mol dm⁻³ HCl

b 0.11 mol dm⁻³ LiOH

c 0.11 mol dm⁻³ Ba(OH)₂

Answers
Acid H₂SO₄: its conjugate base HSO₄⁻
Base CH₃COOH: its conjugate acid CH₃COOH₂⁺
a 0.96
b 13.04
c 13.34

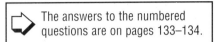

The answers to the numbered questions are on pages 133–134.

1 Calculate the pH of 0.22 mol dm^{-3} C_2H_5COOH which has pK_a = 4.87.

2 25 cm^3 of a weak acid HX of concentration 0.10 mol dm^{-3} was titrated with 0.10 mol dm^{-3} sodium hydroxide solution, and the pH measured at intervals. The results are set out below:

Volume NaOH/cm^3	5	10	12	20	23	24	25	26	30
pH	4.5	4.8	4.9	5.5	6.5	7.0	9.0	12.0	12.5

  a Draw the titration curve and use it to calculate pK_a for the acid HX.

  b Suggest a suitable indicator for the titration.

3 **a** Define a buffer solution and give the name of two substances that act as a buffer when in solution.

  b Explain how this buffer would resist changes in pH, if small amounts of H$^+$ or OH$^-$ ions were added.

4 Calculate the pH of a solution made by adding 4.4 g of sodium ethanoate, CH_3COONa, to 100 cm^3 of a 0.44 mol dm^{-3} solution of ethanoic acid. K_a for ethanoic acid = 1.74×10^{-5} mol dm^{-3}.

Topic **4.5** Organic chemistry II

 Things to learn and understand

Isomerism

❑ **Structural**. There are three types:

  a Carbon chain. Here the isomers have different arrangements of carbon atoms in a molecule,

    e.g. butane, $CH_3CH_2CH_2CH_3$, and methyl propane, $CH_3CH(CH_3)CH_3$.

  b Positional. Here a functional group is on one of two or more different places in a given carbon chain,

    e.g. propan-1-ol, $CH_3CH_2CH_2OH$, and propan-2-ol, $CH_3CH(OH)CH_3$.

  c Functional group. Here the isomers have different functional groups,

    e.g. ethanoic acid, CH_3COOH and methyl methanoate, $HCOOCH_3$.

❑ **Stereoisomerism**. There are two types:

  a **Geometric**. This is caused by having two different groups on each carbon atom of a >C=C< group (see Figure 4.4).

The two isomers are not interconvertible at ordinary temperatures because there is no rotation about a double bond.

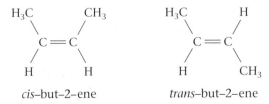

cis–but–2–ene   *trans*–but–2–ene

Fig 4.4 cis *and* trans *isomers of but-2-ene*

Having four different groups attached to one carbon atom will cause this. This carbon atom is called a **chiral** centre.

b Optical. Optical isomers are defined as isomers of which one is the non-superimposable mirror image of the other.

Thus lactic acid (2-hydroxypropanoic acid), $CH_3CH(OH)COOH$, exists as two optical isomers. These must be drawn three dimensionally so as to show that one is the mirror image of the other (see Figure 4.5).

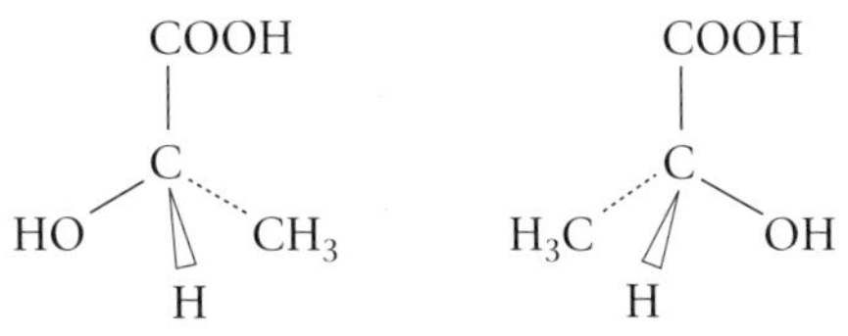

Fig 4.5 *Optical isomers of lactic acid*

A 50/50 mixture of the two isomers will have no effect on plane polarised light. This type of mixture is called a racemic mixture and is often the result when chiral substances are produced by a chemical reaction.

One isomer will rotate the plane of polarisation of plane polarised monochromatic light clockwise and the other will rotate it anti-clockwise.

Further reactions

Grignard reagents

These have the formula R-Mg-halogen such as C_2H_5MgI. They are used to increase the carbon chain length, because they are **nucleophiles** and contain a δ^- carbon atom which will attack a δ^+ carbon atom in other compounds.

Preparation
Halogenoalkane + magnesium
Conditions: dry ether solvent (flammable, therefore no flames), under reflux with a trace of iodine as catalyst: e.g.
$$C_2H_5I + Mg \rightarrow C_2H_5MgI$$

Methanal produces a primary alcohol.

Reactions
- + aldehydes to produce (after hydrolysis with dilute hydrochloric acid) a **secondary alcohol**:
$$C_2H_5MgI + CH_3CHO \rightarrow CH_3CH(OH)C_2H_5$$
butan-2-ol

- + ketones to produce (after hydrolysis) a **tertiary alcohol**:
$$C_2H_5MgI + CH_3COCH_3 \rightarrow CH_3C(CH_3)(OH)C_2H_5$$
2-methyl butan-2-ol

- + solid carbon dioxide (dry ice) to form (after hydrolysis) a **carboxylic acid**:
$$C_2H_5MgI + CO_2(s) \rightarrow C_2H_5COOH$$
propanoic acid

- Grignard reagents react with water to form alkanes:
$$C_2H_5MgI + H_2O \rightarrow C_2H_6$$
which is why they must be prepared and used in dry conditions.

Carboxylic acids

These have the functional group:

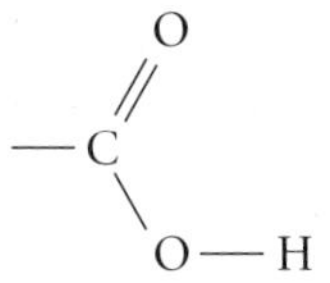

Preparation

They can be made by the oxidation of a primary alcohol. The alcohol is heated under reflux with dilute sulphuric acid and excess potassium dichromate(VI).

Reactions

● + alcohols to produce esters:
$$CH_3COOH + C_2H_5OH \rightleftharpoons CH_3COOC_2H_5 + H_2O$$
ethyl ethanoate
conditions: heat under reflux with a few drops of concentrated sulphuric acid.

● + lithium tetrahydridoaluminate(III) to produce a primary alcohol:
$$CH_3COOH + 4[H] \rightarrow CH_3CH_2OH + H_2O$$
ethanol
conditions: dissolve in dry ether, followed by hydrolysis with $H^+(aq)$.

● + phosphorus pentachloride to produce an acid chloride:
$$CH_3COOH + PCl_5 \rightarrow CH_3COCl + HCl + POCl_3$$
ethanoyl chloride
conditions: dry; observation: steamy fumes given off.

> Test for a carboxylic acid. When added to aqueous sodium hydrogencarbonate (or sodium carbonate), it gives a gas, CO_2, which turns lime water milky.

Esters

These have the general formula RCOOR′ where R and R′ are alkyl or aryl groups, and which may or may not be different.

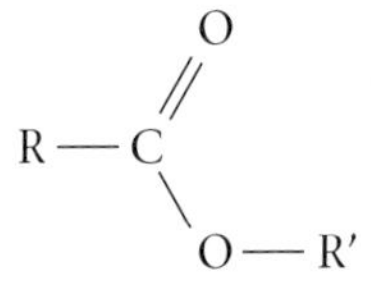

Reactions of ethyl ethanoate

● Hydrolysis with aqueous acid to produce the organic acid and the alcohol:
$$CH_3COOC_2H_5 + H_2O \rightleftharpoons CH_3COOH + C_2H_5OH$$
ethanoic acid
conditions: heat under reflux with dilute sulphuric acid.

● Hydrolysis with aqueous alkali to produce the salt of the acid and the alcohol:
$$CH_3COOC_2H_5 + NaOH \rightarrow CH_3COONa + C_2H_5OH$$
sodium ethanoate
conditions: heat under reflux with aqueous sodium hydroxide.

> *Unit 4*

> *Helpful hints*
>
> The acid is a catalyst. This reaction has a low yield because it is a reversible reaction.

> *Helpful hints*
>
> This reaction has a high yield because it is **not** reversible.

Carbonyl compounds (aldehydes and ketones)

These contain the C=O functional group. Aldehydes have the general formula RCHO, and ketones the general formula RCOR′ where R and R′ are alkyl or aryl groups, and may or may not be different.

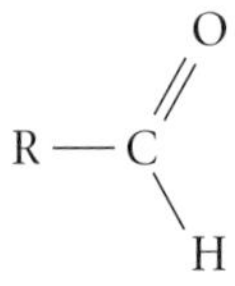

Aldehyde 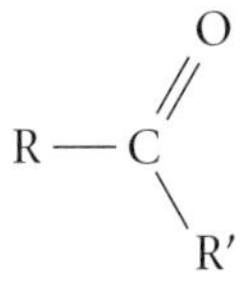   Ketone

Reactions in common

- Addition of hydrocyanic acid, HCN, to produce a hydroxynitrile:

$$CH_3CHO + HCN \rightarrow CH_3CH(OH)CN$$
2-hydroxypropanenitrile

conditions: there must be both HCN and CN$^-$ present,
e.g. HCN + a trace of KOH, or KCN(s) + some dilute sulphuric acid.

- Reduction with lithium tetrahydridoaluminate(III) to produce an alcohol:

$$CH_3CHO + 2[H] \rightarrow CH_3CH_2OH$$
$$CH_3COCH_3 + 2[H] \rightarrow CH_3CH(OH)CH_3$$

conditions: dissolve in dry ether, followed by hydrolysis with H$^+$(aq).
Or reduction with sodium tetrahydridoborate(III), also to produce an alcohol:

$$CH_3CHO + 2[H] \rightarrow CH_3CH_2OH$$
$$CH_3COCH_3 + 2[H] \rightarrow CH_3CH(OH)CH_3$$

conditions: aqueous solution, followed by hydrolysis with H$^+$(aq).

- Reaction with 2,4-dinitrophenylhydrazine (2,4-DNP) to give an orange/red precipitate:

> The formation of an orange or red precipitate with 2,4-DNP is a test for carbonyl compounds, i.e. for both aldehydes and ketones.

2,4-DNP

Specific reaction of aldehydes

Aldehydes are oxidised to carboxylic acids by a solution of silver nitrate in dilute ammonia. On warming the silver ions are reduced to a mirror of metallic silver.
Aldehydes are also oxidised by warming with Fehling's solution, which is reduced from a blue solution to a red precipitate of copper(I) oxide:

$$CH_3CHO + [O] \rightarrow CH_3COOH$$

> The formation of the silver mirror with ammoniacal silver nitrate is the test which distinguishes aldehydes, which give a positive result, from ketones, which have **no** reaction. Aldehydes are also oxidised by potassium manganate(VII) or acidified potassium dichromate(VI).

> The only aldehyde to do this is ethanal, CH$_3$CHO.

Iodoform reaction

- This is a reaction which produces a pale yellow precipitate of iodoform, CHI$_3$, when an organic compound is added to a mixture of iodine and dilute sodium hydroxide (or a mixture of KI and NaOCl).
- The group responsible for this reaction is the CH$_3$C=O group, and the reaction involves breaking the C–C bond between the CH$_3$ group and the C=O group.

$$CH_3COC_2H_5 \text{ produces } C_2H_5COONa + CHI_3(s)$$

- All ketones containing the CH$_3$C=O group will give this precipitate.
- This reaction is also undergone by alcohols containing the CH$_3$CH(OH) group. They are oxidised under the reaction conditions to the CH$_3$C=O group, which then reacts to give iodoform.

Acid chlorides

These have the general formula RCOCl, and the functional group is:

e.g. ethanoyl chloride, CH$_3$COCl, and propanoyl chloride, C$_2$H$_5$COCl.

Helpful hints

This reaction is often used in organic 'problem' questions. The pale yellow precipitate produced on addition of sodium hydroxide and iodine is the clue. The organic compound that produces this precipitate will either be a carbonyl with a CH$_3$C=O group or an alcohol with a CH$_3$CH(OH) group. It is **not** a general test for ketones.

Reactions of ethanoyl chloride

● + water to produce the carboxylic acid, ethanoic acid:
$$CH_3COCl + H_2O \rightarrow CH_3COOH + HCl(g)$$
observation: steamy fumes of HCl.

● + alcohols to produce an ester:
$$CH_3COCl + C_2H_5OH \rightarrow CH_3COOC_2H_5 + HCl$$
observation: characteristic smell of the ester, ethyl ethanoate.

● + ammonia to produce the amide, ethanamide:
$$CH_3COCl + 2NH_3 \rightarrow CH_3CONH_2 + NH_4Cl$$

● + primary amines to produce a substituted amide:
$$CH_3COCl + C_2H_5NH_2 \rightarrow CH_3CONH(C_2H_5) + HCl$$

Nitrogen compounds

Primary amines

● They contain the $-NH_2$ group, e.g. ethylamine $C_2H_5NH_2$.
● They are soluble in water (if the carbon chain is fairly short) because they form hydrogen bonds with water molecules.
● They are weak bases.

Reactions of ethylamine

● + acids, to form an ionic salt that is soluble in water:
$$C_2H_5NH_2 + H^+(aq) \rightleftharpoons C_2H_5NH_3^+(aq)$$

● + acid chlorides, to form a substituted amide:
$$C_2H_5NH_2 + CH_3COCl \rightarrow CH_3CONHC_2H_5 + HCl$$

Nitriles

They contain the $C\equiv N$ group, e.g. propanenitrile C_2H_5CN.

Reations of propanenitrile

● Hydrolysis either with acid:
$$C_2H_5CN + H^+ + 2H_2O \rightarrow C_2H_5COOH + NH_4^+$$
conditions: heat under reflux with dilute sulphuric acid or with alkali.
$$C_2H_5CN + OH^- + H_2O \rightarrow C_2H_5COO^- + NH_3$$
conditions: heat under reflux with dilute sodium hydroxide

● Reduction by lithium tetrahydridoaluminate(III):
$$C_2H_5CN + 4[H] \rightarrow C_2H_5CH_2NH_2$$
conditions: dry ether, followed by addition of dilute acid

Amides

They contain the $CONH_2$ group, e.g. ethanamide CH_3CONH_2.

$$CH_3 - \overset{\displaystyle O}{\underset{\displaystyle NH_2}{\overset{\displaystyle \|}{C}}}$$

Reactions of ethanamide

● Dehydration with phosphorus(V) oxide:
$$CH_3CONH_2 - H_2O \rightarrow CH_3CN$$
conditions: warm and distil off the ethanenitrile.

Helpful hints

The amine can be regenerated from the salt by adding a strong alkali.

Helpful hints

Nitriles can be prepared by the reaction of a halogenoalkane with KCN in aqueous ethanol. Hydroxynitriles can be prepared by the addition of HCN to carbonyl compounds in the presence of base.

Unit 4

● Hofmann degradation reaction:

$$CH_3CONH_2 + Br_2 + 2NaOH \rightarrow CH_3NH_2 + CO_2 + 2NaBr + 2H_2O$$

conditions: add liquid bromine to the amide at room temperature, and then add concentrated aqueous sodium hydroxide solution and warm. The amine distils off.

Amino acids

These contain an NH_2 and a COOH group (usually bonded to the same carbon atom). They react both as acids and bases.

They are water-soluble ionic solids because they form a **zwitterion**.

$$NH_2CH_2COOH \rightleftharpoons NH_3^+CH_2COO^-$$

Reactions

● With acids:

$$NH_2CH_2COOH + H^+(aq) \rightarrow NH_3^+CH_2COOH(aq)$$

● With bases:

$$NH_2CH_2COOH + OH^-(aq) \rightarrow NH_2CH_2COO^-(aq) + H_2O$$

Figure 4.6 summarises the reactions of organic substances.

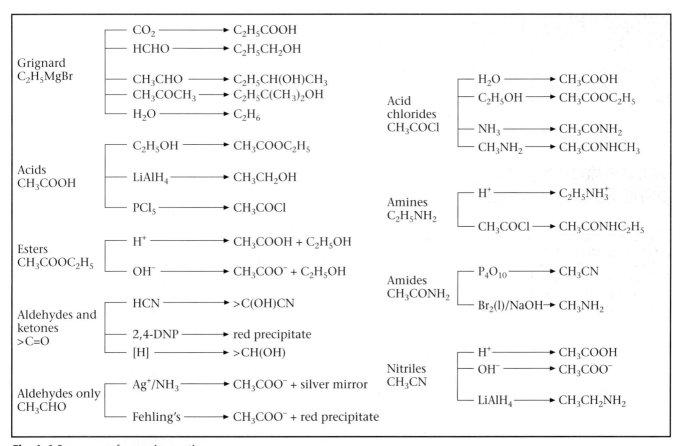

Fig 4.6 *Summary of organic reactions*

Checklist

Before attempting questions on this topic, check that you:

☐ Can recognise stereoisomerism (geometric and optical) in organic compounds.

☐ Know the effect of an optical isomer on plane polarised light.

☐ Understand the nature of a racemic mixture.

☐ Can recall the preparation and reactions of Grignard reagents.

☐ Can recall the reactions of carboxylic acids.

☐ Can recall the reactions of esters.

☐ Can recall the reactions of carbonyl compounds (aldehydes and ketones).

☐ Can recall the reactions of ethanoyl chloride.

☐ Can recall the reactions of amines with acids and with acid chlorides.

☐ Can recall the hydrolysis and the reduction of nitriles.

☐ Can recall the reactions of amides with P_4O_{10} and with $Br_2/NaOH$.

☐ Know that amino acids are both acids and bases.

 Testing your knowledge and understanding

For the following questions, cover the margin, write your answer, then check to see if you are correct.

● Write the formulae of the products of the reaction of CH_3MgI with:
  a propanal
  b butan-2-one
  c methanal.

● State the names of the reagents and give the conditions for the following conversions:
  a $C_2H_5COOCH_3$ to $C_2H_5COOH + CH_3OH$ in a high yield
  b C_2H_5COOH to $CH_3CH_2CH_2OH$.

● State the names of the reagents needed to convert ethanoyl chloride to:
  a ethanoic acid
  b methyl ethanoate
  c ethanamide.

● State the structural formulae of the product obtained by reacting:
  a ethylamine with dilute hydrochloric acid
  b ethylamine with ethanoyl chloride.

Answers

a	$CH_3CH_2CH(OH)CH_3$
b	$CH_3CH_2C(OH)(CH_3)CH_3$
c	CH_3CH_2OH

a	Heat under reflux with dilute sodium hydroxide, then add dilute sulphuric acid.
b	Add lithium tetrahydridoaluminate(III) in dry ether, followed by dilute sulphuric acid.

a	water
b	methanol
c	ammonia

a	$CH_3CH_2NH_3^+\ Cl^-$
b	$CH_3CONHCH_2CH_3$

Unit 4

a	Heat under reflux with NaOH(aq) then acidify.
b	Add LiAlH₄ in dry ether, followed by dilute alkali.
c	Warm with liquid bromine and concentrated aqueous sodium hydroxide.

Because it forms the zwitterion $^+NH_3CH_2COO^-$

 The answers to the numbered questions are on page 134.

● State the conditions for the conversion of:
  a ethanenitrile, CH_3CN, to ethanoic acid, CH_3COOH
  b ethanenitrile, CH_3CN, to ethylamine, $C_2H_5NH_2$
  c propanamide, $C_2H_5CONH_2$, to ethylamine, $C_2H_5NH_2$.

● Explain why aminoethanoic acid is a solid that is soluble in water.

1 Draw the stereoisomers of:
  a $CH_3CH=C(Cl)CH_3$
  b $CH_3CH(OH)F$.
2 Outline how you would prepare 2-methylpropanoic acid from 2-iodopropane.
3 Describe the tests that you could do to distinguish between:
  a ethanoic acid and ethanoyl chloride
  b propanal and propanone.

Unit 4

Practice Test: Unit 4

Time allowed 1 hr 30 min

All questions are taken from parts of previous Edexcel Advanced GCE papers

The answers are on pages 134–136.

1 a Draw a Born–Haber cycle for the formation of magnesium chloride, $MgCl_2$, and use the values below to calculate the lattice energy of magnesium chloride. **[5]**

	ΔH°/kJ mol^{-1}
1st electron affinity of chlorine	–364
1st ionisation energy of magnesium	+736
2nd ionisation energy of magnesium	+1450
Enthalpy of atomisation of chlorine	+121
Enthalpy of atomisation of magnesium	+150
Enthalpy of formation of $MgCl_2(s)$	–642

b Use the following to answer the questions in this section.

	ΔH°/kJ mol^{-1}
Enthalpy of hydration of Sr^{2+}	–1480
Enthalpy of hydration of Ba^{2+}	–1360
Enthalpy of hydration of OH^-	–460
Lattice energy of $Sr(OH)_2(s)$	–1894
Lattice energy of $Ba(OH)_2(s)$	–1768

  i Explain why the lattice energy of strontium hydroxide is different from that of barium hydroxide. **[2]**
  ii Explain why the hydration enthalpy of a cation is exothermic. **[2]**
  iii Use the lattice energy and hydration enthalpy values to explain why barium hydroxide is more soluble in water than strontium hydroxide. **[4]**

(Total 13 marks)
[January 2001 CH 3 question 4]

2 a i Give the structural formula of a nitrile, C_4H_7N, that has an unbranched chain **[1]**
  ii Primary amines can be made by reducing nitriles. Suggest a reagent that could be used for this purpose. **[1]**
  iii Draw the structural formula of the amine produced by reducing the nitrile given in **a** part **i**. **[1]**
 b i What feature of an amine molecule makes it both a base and a nucleophile? **[1]**
  ii Give, by writing an equation, an example of an amine acting as a base **[1]**
 c Ethanoyl chloride, CH_3COCl, reacts with both amines and alcohols
  i Give the full structural formula of the compound produced when ethanoyl chloride reacts with ethylamine, $C_2H_5NH_2$. **[1]**
  ii Name the **type** of the compound produced in **c** part **i**. **[1]**
  iii State **one** of the advantages of reacting ethanoyl chloride with ethanol to make an ester rather than reacting ethanoic acid with ethanol. **[1]**
  iv Write the full structural formula of the ester made in **c** part **iii** **[1]**

(Total 9 marks)
[January 2002 Unit Test 4 question 5 - modified]

3 Compound **X** is a secondary alcohol which has a molecular formula of $C_4H_{10}O$.
   X exhibits optical isomerism.
   a Draw the structural formulae of the two optical isomers of **X**. [2]
   b Suggest reagents and conditions which would enable its preparation via a reaction involving a
     Grignard reagent. (You are not expected to describe how you would prepare a Grignard reagent) [3]
   c Compound **Y** is obtained by oxidising the secondary alcohol **X** with potassium dichromate(VI)
     acidified with dilute sulphuric acid.
       i Draw the structural formula of **Y**. [1]
       ii To which class of compounds does **Y** belong? [1]
       iii Describe the tests you would do on **Y**, and the results you would expect, to show that your
           classification is correct. [4]
       iv Both **X** and **Y** give a yellow precipitate when treated with iodine in the presence of sodium
           hydroxide solution. Write the structural formulae of the organic products of this reaction. [2]

(Total 13 marks)
[June 2001 CH 4 question 1- modified]

4   At about 1000 °C, when aluminium chloride vapour is heated with solid aluminium, the following
    equilibrium is set up:

$$AlCl_3(g) + 2Al(s) \rightleftharpoons 3AlCl(g)$$

   a Give the expression for the equilibrium constant, K_p, for this reaction. [1]
   b At a particular temperature, a mixture of the above system at equilibrium was found to contain 0.67 g
     of $AlCl_3$ and 0.63 g of AlCl vapours at an equilibrium pressure of 2.0 atm. Calculate the value of the
     equilibrium constant, K_p, at this temperature, stating its units. [4]
   c The position of equilibrium moves to the right as the temperature is raised. State what this suggests
     about the enthalpy change for the forward reaction. [1]
   d State, with reasoning, the effect of an increase in the pressure on the system on the value of the
     equilibrium constant and on the position of the equilibrium. [3]
   e Suggest how this reversible reaction could be used in a process to recycle aluminium by extracting it
     from impure aluminium. [2]

(Total 11 marks)
[June 2001 CH3 question 3]

5 a i Write an equation for the reaction between magnesium oxide and dilute sulphuric acid, including
        the state symbols. [2]
     ii Describe what you would see during this reaction. [2]
   b i Write an equation for the reaction between phosphorus(V) oxide and aqueous sodium hydroxide
        solution. [2]
     ii With the aid of two equations, show how aluminium hydroxide exhibits amphoteric behaviour. [3]

   c   With reference to the reactions in **a** and **b**, describe the variation in the metallic character of the
       elements across Period 3 of the Periodic Table (sodium to argon) [2]
   d   Suggest, with reasoning, the acid-base character of Indium(III) oxide, In_2O_3. Indium is the fourth
       element down Group 3 of the Periodic Table. [2]

(Total 13 marks)
[June 2002 Unit Test 4 question 4]

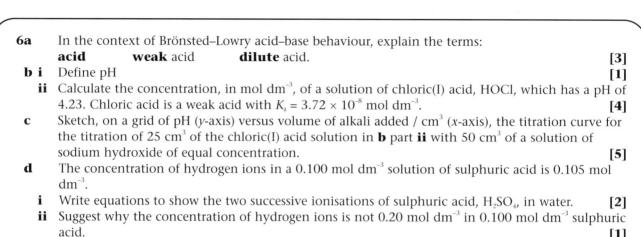

6a In the context of Brönsted–Lowry acid–base behaviour, explain the terms:
acid   **weak** acid   **dilute** acid. [3]
b i Define pH [1]
ii Calculate the concentration, in mol dm^{-3}, of a solution of chloric(I) acid, HOCl, which has a pH of 4.23. Chloric acid is a weak acid with $K_a = 3.72 \times 10^{-8}$ mol dm^{-3}. [4]
c Sketch, on a grid of pH (*y*-axis) versus volume of alkali added / cm^3 (*x*-axis), the titration curve for the titration of 25 cm^3 of the chloric(I) acid solution in **b** part **ii** with 50 cm^3 of a solution of sodium hydroxide of equal concentration. [5]
d The concentration of hydrogen ions in a 0.100 mol dm^{-3} solution of sulphuric acid is 0.105 mol dm^{-3}.
i Write equations to show the two successive ionisations of sulphuric acid, H$_2$SO$_4$, in water. [2]
ii Suggest why the concentration of hydrogen ions is not 0.20 mol dm^{-3} in 0.100 mol dm^{-3} sulphuric acid. [1]

(Total 16 marks)

[January 2001 CH2 question 4 & January 2002 CH2 question 2]

5 Transition metals, quantitative kinetics and applied organic chemistry

Topic 5.1 Redox equilibria

 ### Introduction

- You must revise Topic 1.5 (Introduction to oxidation and reduction).
- The sign of $E^\ominus$ indicates the direction of spontaneous reaction.
- A useful mnemonic is OIL RIG. When a substance is **O**xidised **It** **L**oses one or more electrons, and if **R**educed **It** **G**ains one or more electrons.
- A half equation always contains electrons.
- Half equation data are usually given as reduction potentials, i.e. with the electrons on the **left**-hand side.
- An oxidising agent becomes reduced when it reacts.

 ### Things to learn and understand

The standard hydrogen electrode

This consists of hydrogen gas at 1 atm pressure bubbling over a platinum electrode immersed in a 1 mol dm^{-3} solution of H$^+$ ions, at a temperature of 25 °C. By definition the potential of a standard hydrogen electrode is zero (see Figure 5.1).

Standard electrode potential, $E^\ominus$

- This is the electric potential (EMF) of a cell composed of the standard electrode connected to a standard hydrogen electrode, measured when the concentrations of all the ions are 1 mol dm^{-3}, the temperature is 25 °C and any gases are at 1 atm pressure.

  - **for a metal** the standard electrode potential is when the metal is immersed in a solution of its ions at a concentration of 1 mol dm^{-3}.

    For the reaction:

    $$Zn^{2+}(aq) + 2e^- \rightarrow Zn(s)$$

    it is for a zinc electrode immersed in a 1 mol dm^{-3} solution of Zn^{2+} ions at 25 °C.

  - **for a non-metal** the standard electrode potential is when the non-metal at 1 atm pressure (or a solution of it at a concentration of 1 mol dm^{-3}) is in contact with a platinum electrode immersed in a

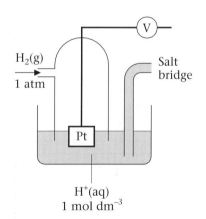

Fig 5.1 A standard hydrogen electrode

solution of the non-metal's ions at a concentration of 1 mol dm^{-3} at 25 °C.

For the reaction:

$$\tfrac{1}{2}Cl_2(g) + e^- \rightarrow Cl^-(aq)$$

it is for chlorine gas, at 1 atm pressure, being bubbled over a platinum electrode dipping into a 1 mol dm^{-3} solution of Cl$^-$ ions at 25 °C.

● **for a redox system of two ions of an element** the standard electrode potential is when a platinum electrode is immersed in a solution containing all the ions in the half equation at a concentration of 1 mol dm^{-3}.

For the reaction:

$$Fe^{3+}(aq) + e^- \rightarrow Fe^{2+}(aq)$$

it is for a platinum electrode immersed in a solution which is 1 mol dm^{-3} in both Fe^{3+} and Fe^{2+} ions at 25 °C.

● The equations are always given as **reduction** potentials, i.e. with the electrons on the **left**.

● Comparing substances on the **left** of two half equations, the one with the **larger** positive value of $E^{\ominus}$ is the more powerful oxidising agent (it is the most easily reduced).

● Comparing substances on the **right** of two half equations, the one with the **smaller** positive (or more **negative**) value of $E^{\ominus}$ is the more powerful reducing agent.

Calculation of $E_{reaction}$ (E_{cell})

The value can be deduced in one of three ways:

1 From a cell diagram:

$$E_{cell} = E_{\text{right-hand electrode}} - E_{\text{left-hand electrode}}$$

If E_{cell} works out to be a negative value, you have written the cell diagram backwards, and so the reaction will go right to left.

2 From an overall equation: write both half reactions as reduction potentials, i.e. with electrons on the left.

$E_{reaction}$ is calculated as:

> $E_{reaction}$ = (E of half equation of the reactant being reduced) – (E of half equation of reactant being oxidised.

Thus for the reaction:

$$Zn(s) + Cu^{2+}(aq) \rightleftharpoons Zn^{2+}(aq) + Cu(s)$$

The half equations are:

$$Cu^{2+}(aq) + 2e^- \rightleftharpoons Cu(s) \qquad E^{\ominus} = +0.34 \text{ V}$$
$$Zn^{2+}(aq) + 2e^- \rightleftharpoons Zn(s) \qquad E^{\ominus} = -0.76 \text{ V}$$

In the reaction the Cu^{2+} ions are being reduced (electron gain), and the zinc atoms are being oxidised (electron loss):

$$E^{\ominus}_{reaction} = +0.34 - (-0.76) = +1.10 \text{ V}$$

3 From half equations:

To do this one of the half equations has to be reversed (see Worked example below). This process alters the sign of its $E^{\ominus}$.

Helpful hints

$\tfrac{1}{2}Cl_2 + e^- \rightleftharpoons Cl^- \; E^{\ominus} = +1.36$ V
$\tfrac{1}{2}Br_2 + e^- \rightleftharpoons Br^- \; E^{\ominus} = +1.07$ V
As +1.36 > +1.07, chlorine is a stronger oxidising agent than bromine.

Helpful hints

$Sn^{4+} + 2e^- \rightleftharpoons Sn^{2+} \; E^{\ominus} = +0.15$ V
$Fe^{3+} + e^- \rightleftharpoons Fe^{2+} \qquad E^{\ominus} = +0.77$ V
As +0.15 < +0.77, Sn^{2+} ions are a better reducing agent than Fe^{2+} ions.

Helpful hints

When a redox half equation is reversed, its sign must be changed. The number of electrons on the left-hand side of one half equation must equal the number on the right-hand side of the other half equation. When a redox half equation is multiplied, its $E^{\ominus}$ value is **not** altered.

Always ensure that the **reactants** that are specified in the question (here they are iron(II) and manganate(VII) ions) are on the **left**-hand side of the final overall equation.

Helpful hints

In the reaction above, the oxidation number of the manganese decreases by 5 from +7 to +2, and as the iron increases by 1, there must be 5 iron(II) ions to each MnO_4^- ion in the overall equation.

Worked example

Use the following data to deduce the overall equation and the value of $E^{\ominus}_{reaction}$ for the reaction between acidified potassium manganate(VII) and iron (II) ions.

i $MnO_4^-(aq) + 8H^+(aq) + 5e^- \rightleftharpoons Mn^{2+}(aq) + 4H_2O(aq)$ $E^{\ominus} = +1.52$ V

ii $Fe^{3+}(aq) + e^- \rightleftharpoons Fe^{2+}(aq)$ $E^{\ominus} = +0.77$ V

Answer.
Reverse equation (ii) and multiply it by 5. Then add it to equation (i):

$5Fe^{2+}(aq) \rightleftharpoons 5Fe^{3+}(aq) + 5e^-$ $E^{\ominus} = -(+0.77$ V$) = -0.77$ V

$MnO_4^-(aq) + 8H^+(aq) + 5e^- \rightleftharpoons Mn^{2+}(aq) + 4H_2O(l)$ $E^{\ominus} = +1.52$ V

$MnO_4^-(aq) + 8H^+(aq) + 5Fe^{2+}(aq) \rightleftharpoons Mn^{2+}(aq) + 4H_2O(l) + 5Fe^{3+}(aq)$

$E^{\ominus}_{reaction} = -0.77 + 1.52 = +0.75$ V

Stoichiometry of an overall equation

The total increase in oxidation number of one element must equal the total decrease in another.

Spontaneous change

- Electrode potential data can be used to predict the feasibility of a chemical reaction. A reaction is feasible (thermodynamically unstable) if E_{cell} is positive.
- However the rate of the reaction may be so slow that the reaction is not observed (kinetically stable).

Non-standard cells

- Non-standard conditions may result in a reaction taking place even if the **standard** electrode potential is negative.
- The reaction:

$$2Cu^{2+}(aq) + 4I^-(aq) \rightleftharpoons 2CuI(s) + I_2(aq)$$

should not work because $E^{\ominus}$ (assuming all species are soluble) $= -0.39$ V. However copper(I) iodide is precipitated and this makes $[Cu^+(aq)]$ very much less than 1 mol dm^{-3}. The equilibrium is driven to the right by the removal of Cu^+ (aq) ions, so that $E_{reaction}$ becomes positive and the reaction takes place.

Potassium manganate(VII) titrations – estimation of reducing agents

- Acidified potassium manganate(VII) will quantitatively oxidise many reducing agents.
- The procedure is to pipette a known volume of the reducing agent into a conical flask and add an excess of dilute sulphuric acid.
- The potassium manganate(VII) solution of known concentration is put in the burette and run in **until there is a faint pink colour**.
- This shows that there is a minute excess of the manganate(VII) ions.
- If the stoichiometry of the reaction is known, the concentration of the reducing agent can be calculated. No indicator is needed as the manganate(VII) ions are very intensely coloured.

Worked example

25.0 cm³ of a solution of iron(II) sulphate was acidified, and titrated against 0.0222 mol dm⁻³ potassium manganate(VII) solution. 23.4 cm³ were required to give a faint pink colour. Calculate the concentration of the iron(II) sulphate solution.

Answer: The equation for the reaction is:

$$MnO_4^-(aq) + 8H^+(aq) + 5Fe^{2+}(aq) \rightarrow Mn^{2+}(aq) + 4H_2O(l) + 5Fe^{3+}(aq)$$

Amount of manganate(VII) = $0.0222 \times 23.4/1000 = 5.195 \times 10^{-4}$ mol

Amount of iron(II) sulphate = $5.195 \times 10^{-4} \times 5/1 = 2.597 \times 10^{-3}$ mol

Concentration of iron(II) sulphate = $2.597 \times 10^{-3} \div 0.0250 = 0.104$ mol dm⁻³

> The ratio of iron(II) to manganate(VII) ions in the equation is 5:1 and so the number of moles of Fe^{2+} is 5 times the number of moles of manganate(VII).

Iodine/thiosulphate titrations – estimation of oxidising agents

- The procedure is to add a 25.0 cm³ sample of an oxidising agent to excess potassium iodide solution (often in the presence of dilute sulphuric acid).
- The oxidising agent liberates iodine, which can then be titrated against standard sodium thiosulphate solution. When the iodine has faded to a pale straw colour, starch indicator is added, and the addition of sodium thiosulphate continued until the blue colour disappears.
- Iodine reacts with thiosulphate ions according to the equation:

$$I_2 + 2S_2O_3^{2-} \rightarrow 2I^- + S_4O_6^{2-}$$

Worked example

25.0 cm³ of a solution of hydrogen peroxide, H_2O_2, was added to an excess of acidified potassium iodide solution, and the liberated iodine required 23.8 cm³ of 0.106 mol dm⁻³ sodium thiosulphate solution. Calculate the concentration of the hydrogen peroxide solution.

Answer. The equation for the oxidation of iodide ions by hydrogen peroxide is:

$$H_2O_2 + 2H^+ + 2I^- \rightarrow I_2 + 2H_2O$$

Amount of sodium thiosulphate	= $0.106 \times 23.8/1000$	= 2.523×10^{-3} mol
Amount of iodine produced	= $2.523 \times 10^{-3} \times 1/2$	= 1.261×10^{-3} mol
Amount of hydrogen peroxide	= $1.261 \times 10^{-3} \times 1/1$	= 1.261×10^{-3} mol
Concentration of H_2O_2	= $1.261 \times 10^{-3} \div 0.0250$	= 0.0505 mol dm⁻³

> The ratio of iodine to thiosulphate ions is 1:2, so the moles of iodine are $\frac{1}{2}$ the moles of thiosulphate.

Disproportionation

A disproportionation reaction can be predicted by using electrode potentials. Use the data to work out E_{cell} for the proposed disproportionation reaction. If it is positive, the reaction will occur.

Worked example

Will copper(I) ions disproportionate in aqueous solution to copper and copper(II) ions?

$$Cu^+(aq) + e^- \rightleftharpoons Cu(s) \qquad E^{\ominus} = +0.52 \text{ V}$$
$$Cu^{2+}(aq) + e^- \rightleftharpoons Cu^+(aq) \qquad E^{\ominus} = +0.15 \text{ V}$$

Answer. Reverse the second equation and add it to the first. This gives the equation:

$$2Cu^+(aq) \rightleftharpoons Cu^{2+}(aq) + Cu(s) \qquad E^{\ominus}_{cell} = +0.52 - (+0.15) = +0.37 \text{ V}$$

Because $E^{\ominus}_{cell}$ is positive, the reaction is feasible, and so aqueous copper(I) ions will disproportionate.

Corrosion (rusting)

This is an electrolytic process. When iron is stressed or pitted, some areas become anodic and some cathodic. In the presence of water and oxygen the following reactions take place (see Figure 5.2).

● At the anodic areas iron atoms become oxidised and lose 2 electrons:

$$Fe(s) \rightarrow Fe^{2+}(aq) + 2e^-$$

● The electrons travel through the metal and reduce oxygen :

$$\tfrac{1}{2}O_2(aq) + 2e^- + H_2O \rightarrow 2OH^-(aq)$$

● The Fe^{2+} and the OH^- ions meet and iron(II) hydroxide is precipitated:

$$Fe^{2+}(aq) + 2OH^-(aq) \rightarrow Fe(OH)_2(s)$$

● Finally oxygen oxidises the iron(II) hydroxide to iron(III) oxide (rust):

$$2Fe(OH)_2(s) + \tfrac{1}{2}O_2(aq) \rightarrow Fe_2O_3(s) + 2H_2O(l)$$

Fig 5.2 Rusting

Stainless steel is iron containing a high proportion of chromium, which forms a protective layer of Cr_2O_3 over the whole surface of the metal. If the surface is scratched, a new protective layer of Cr_2O_3 is formed.

Prevention of corrosion

This can be done by:
● placing a physical barrier between the steel and the environment. Such barriers are paint, tin or chromium plating.
● adding a sacrificial metal. This can be done by coating with zinc (galvanising), or by attaching blocks of magnesium at intervals.

Storage cells

These store electrical energy as chemical energy. The reactions must be fully reversible and the chemicals produced in the redox reactions must be insoluble.

The lead acid battery

When electricity is drawn from the cell (discharging), the following reactions take place.
At the anode (oxidation) which is negative:

$$Pb(s) + SO_4^{2-}(aq) \rightarrow PbSO_4(s) + 2e^- \qquad E^{\ominus} = +0.36 \text{ V}$$

At the cathode (reduction) which is positive:

$$PbO_2(s) + 2e^- + SO_4^{2-}(aq) + 4H^+(aq) \rightarrow PbSO_4(s) + 2H_2O(l)$$
$$E^{\ominus} = +1.69 \text{ V}$$

The reactions for charging the cell are the opposite, and only take place if a potential >2.05 V is applied, with the anode being connected to the negative terminal of the charging source.

The overall discharging reaction is:

$$Pb(s) + PbO_2(s) + 2SO_4^{2-}(aq) + 4H^+(aq) \rightarrow 2PbSO_4(s) + 2H_2O(l)$$
$$E^{\ominus} = +1.69 + (+0.36) = +2.05 \text{ V}$$

Unit 5

Checklist

Before attempting the questions on this topic, check that you can:

☐ Define standard electrode potential.

☐ Describe the construction of a standard hydrogen electrode.

☐ Write half equations and use them to deduce overall equations.

☐ Predict the feasibility of redox and disproportionation reactions.

☐ Deduce oxidation numbers and use them to balance redox equations.

☐ Recall the principles of manganate(VII) and thiosulphate titrations.

☐ Recall the corrosion of iron and its prevention.

☐ Understand the chemistry of the lead/acid battery.

Testing your knowledge and understanding

For the following set of questions, cover the margin before you answer, then check to see if you are correct.

● Write the equation representing the reaction that takes place in a standard hydrogen electrode.

● What are the conditions, other than temperature, for this electrode?

● What are the oxidation numbers of manganese in MnO_4^- and in Mn^{2+}?

● What are the oxidation numbers of sulphur in SO_3^{2-} and in SO_4^{2-}?

● What is the ratio of SO_3^{2-} to MnO_4^- in the reaction between them?

● Why is it not necessary to have an indicator present in potassium manganate(VII) titrations?

1 a Write ionic half equations for the reduction of:

   i $Cr_2O_7^{2-}$ to Cr^{3+} in acidic solution $E^\ominus = +1.33$ V

   ii Sn^{4+} to Sn^{2+} $E^\ominus = +0.15$ V

   iii   Iodate(V) ions, (IO_3^-) to I_2 in acidic solution $E^\ominus = +1.19$ V

   iv I_2 to I^- $E^\ominus = +0.54$ V

b Write overall ionic equations, calculate $E_{reaction}$ values and hence comment on the feasibility of the reactions between:

   i   potassium dichromate(VI) and tin(II) chloride in acid solution

   ii   potassium iodate(V) and potassium iodide in acid solution.

2 1.32 g of mild steel filings was reacted with excess dilute sulphuric acid, and the resulting solution made up to a volume of 250 cm³. 25.0 cm³ samples of this were titrated against 0.0200 mol dm⁻³ potassium manganate(VII) solution. The mean titre was 23.5 cm³. Calculate the percentage of iron in the steel.

3 Write the half equations for the redox reactions involved in the corrosion of iron.

Answers

$\frac{1}{2} H_2(g) \rightleftharpoons H^+(aq) + e^-$. State symbols are essential.
H_2 (g) at 1 atm pressure, $[H^+(aq)] = 1$ mol dm⁻³
+7 and +2
+4 and +6
5:2 So that oxidation numbers of both change by 10
Because the manganate(VII) ions are intensely coloured.

 The answers to the numbered questions are on pages 136–137.

Unit 5

Topic 5.2 Transition metal chemistry

Introduction

- ☐ You should know the colour of the aqua complex ions and of the hydroxides of the d block elements scandium to zinc.

- ☐ You must be able to link the reactions in this topic to the theory of redox equilibria, Topic 5.1.

Things to learn

> Neither scandium nor zinc is a transition metal although they are in the d block, because their ions have the electronic structure [Ar] $3d^0$, $4s^0$ and [Ar], $3d^{10}$, $4s^0$ respectively.

- ● **d block elements** are those in which the highest occupied energy level is a d orbital.
- ● A **transition element** is one that has at least one of its ions with a partly filled d shell.
- ● **Electron structure of the atoms**

	Sc	Ti	V	Cr	Mn	Fe	Co	Ni	Cu	Zn
3d	1	2	3	**5**	5	6	7	8	**10**	10
4s	2	2	2	**1**	2	2	2	2	**1**	2

> Note that the number of d electrons increases left to right, except that Cr and Cu have $4s^1$ electron structures, whereas the others have $4s^2$. This is because stability is gained when the d shell is half full or full.

- ● **Electron structure of the ions**. The element first loses its 4s electrons when forming an ion. Thus the electron structures of iron and its ions are:

$$Fe \quad [Ar], 3d^6, 4s^2$$
$$Fe^{2+} \quad [Ar], 3d^6, 4s^0$$
$$Fe^{3+} \quad [Ar], 3d^5, 4s^0$$

Things to understand

Properties of transition metals

Complex ion formation

- ● **Bonding in complex ions**. The simplest view is that the ligands form dative covalent bonds by donating a lone pair of electrons into empty orbitals of the transition metal ion. These could be empty 3d, 4s, 4p or 4d orbitals in the ion (e.g. see the arrangement of electrons in boxes in Figure 5.3).

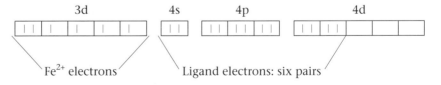

Fig 5.3 *Electron arrangement in the hexa-aquairon(II) ion*

Unit 5

Typical ligands are:

neutral molecules: H_2O (aqua), NH_3 (ammine)

anions: F^- (fluoro), Cl^- (chloro), $(CN)^-$ (cyano).

- Shapes of complex ions (see pages 14–15). If the complex ion has 6 pairs of electrons donated by ligands, the complex will be **octahedral**. This is so that the electron pairs in the ligands will be as far apart from each other as possible.

Coloured complex ions

- If the ion has partially filled d orbitals, it will be coloured. Sc^{3+} and Ti^{4+} have d^0 structures, and Cu^+ and Zn^{2+} have d^{10} and so are not coloured.
- d orbitals point in different directions in space, and so interact to different extents with the electrons in the ligands.
- This causes a splitting of the d orbitals into two of higher energy and three of lower energy.
- When white light is shone into the substance, a d electron is moved from the lower energy to the higher energy level.
- The frequency of the light that causes this jump is in the visible range, so that colour at this frequency is **removed** from the white light.
- The colour depends on the size of the energy gap which varies with the metal ion and with the type of ligand.

> $[Cu(H_2O)_6]^{2+}$ is turquoise blue, $[Cu(NH_3)_4(H_2O)_2]^{2+}$ is deep blue, and $[CuCl_4]^{2-}$ is yellow.

Variable oxidation state

- Successive ionisation energies increase steadily until all the 4s and the 3d electrons have been removed, after which there is a large jump in the value.
- Formation of cations. The increase between successive ionisation energies is compensated for by a similar increase in hydration energies. Thus cations in different oxidation states are energetically favourable for all transition metals.
- Formation of covalent bonds. Transition metals can make available a variable number of d electrons for covalent bonding. The energy required for the promotion of an electron from a 3d to a higher energy orbital is compensated for by the bond energy released.
- Formation of oxoions. In ions such as MnO_4^-, VO_3^-, VO_2^+ and VO^{2+}, the oxygen is covalently bonded to the transition metal which uses a varying number of d electrons.
- Manganese exists:

in the +2 state as Mn^{2+}
in the +4 state as MnO_2
in the +6 state as MnO_4^{2-}
in the +7 state as MnO_4^-.

> Hydrated cations of charge 4+ or more do not exist because the sum of all the ionisation energies is too large to be compensated for by the hydration energy.
> Copper forms Cu^+ and Cu^{2+}, and chromium Cr^{2+} and Cr^{3+}.

Catalytic activity

Transition metals and their compounds are often good catalysts.
- Vanadium(V) oxide is used in the oxidation of SO_2 to SO_3 in the contact process for the manufacture of sulphuric acid.
- Iron is used in the Haber process for the manufacture of ammonia.
- Nickel is used in the addition of hydrogen to alkenes (hardening of vegetable oils).

Reactions of transition metal compounds

Reactions with sodium hydroxide and ammonia solutions

The table gives a summary of these reactions.

Ion [a]	Colour	Addition of NaOH(aq)	Excess NaOH(aq)	Addition of NH$_3$(aq)	Excess NH$_3$(aq)
Cr^{3+}	Green	Green ppt	Green solution	Green ppt	Ppt remains
Mn^{2+}	Pale pink	Sandy ppt[b]	Ppt remains	Sandy ppt[b]	Ppt remains
Fe^{2+}	Pale green	Dirty green ppt[c]	Ppt remains	Dirty green ppt[c]	Ppt remains
Fe^{3+}	Brown/yellow	Foxy red ppt	Ppt remains	Foxy red ppt	Ppt remains
Ni^{2+}	Green	Green ppt	Ppt remains	Green ppt	Blue soln
Cu^{2+}	Pale blue	Blue ppt[d]	Ppt remains	Pale blue ppt	Deep blue soln
Zn^{2+}	None	White ppt	Colourless solution	White ppt	Colourless soln

[a] The correct formula for the ions should be the hexa-aqua ion, except for zinc, which forms a tetraqua ion.
[b] The precipitate of Mn(OH)$_2$ goes brown as it is oxidised by air.
[c] The precipitate of Fe(OH)$_2$ goes brown on the surface as it is oxidised by the oxygen in the air.
[d] The precipitate of Cu(OH)$_2$ goes black as it loses water to form CuO.

Deprotonation

- The aqua ions in solution are partially deprotonated by water. The greater the surface charge density of the ion the greater the extent of this reaction, e.g. hexa-aqua iron(III) ions:

$$[Fe(H_2O)_6]^{3+}(aq) + H_2O \rightleftharpoons [Fe(H_2O)_5(OH)]^{2+}(aq) + H_3O^+(aq)$$

This means that solutions of iron(III) ions are acidic (pH < 7).

- When an alkali such as sodium hydroxide is added, the equilibrium is driven to the right, the ion is considerably deprotonated to form a neutral molecule which loses water to form a precipitate of the metal hydroxide:

$$[Fe(H_2O)_6]^{3+}(aq) + 3OH^-(aq) \rightarrow Fe(OH)_3(s) + 6H_2O$$

If aqueous ammonia is added, the same precipitate is formed:

$$[Fe(H_2O)_6]^{3+}(aq) + 3NH_3(aq) \rightarrow Fe(OH)_3(s) + 3NH_4^+(aq) + 3H_2O$$

Ligand exchange

- When aqueous ammonia is added to aqua complexes of d block elements such as those of nickel, copper and zinc, ligand exchange takes place and a solution of the ammine complex is formed.

First the hydroxide is precipitated in a **deprotonation** reaction:

$$[Cu(H_2O)_6]^{2+}(aq) + 2NH_3(aq) \rightarrow Cu(OH)_2(s) + 2NH_4^+(aq) + 4H_2O$$

The hydroxide then ligand exchanges to form an ammine complex with excess ammonia:

$$Cu(OH)_2(s) + 4NH_3(aq) + 2H_2O \rightarrow [Cu(NH_3)_4(H_2O)_2]^{2+} + 2OH^-(aq)$$

The final result is that the NH$_3$ ligand has taken the place of four H$_2$O ligands.

- Addition of cyanide ions to iron(II) ions produces a solution of hexacyanoferrate(II):

$$[Fe(H_2O)_6]^{2+}(aq) + 6CN^-(aq) \rightarrow [Fe(CN)_6]^{4-}(aq) + 6H_2O$$

- A test for iron(III) ions is to add a solution of potassium thiocyanate, KCNS. Iron(III) ions give a blood red solution:

$$[Fe(H_2O)_6]^{3+}(aq) + SCN^-(aq) \rightarrow [Fe(SCN)(H_2O)_5]^{2+}(aq) + H_2O$$

Amphoteric hydroxides 'redissolve' in excess strong alkali, e.g.
Cr(OH)$_3$(s) + 3OH$^-$(aq) $\rightarrow$ [Cr(OH)$_6$]$^{3-}$(aq)
and Zn(OH)$_2$(s) + 2OH$^-$(aq) $\rightarrow$ [Zn(OH)$_4$]$^{2-}$(aq).

Helpful hints

This reaction is used as a test for:
a nickel(II): with aqueous ammonia nickel(II) ions first give a green precipitate which then forms a blue solution with excess ammonia.
b copper(II) ions give a pale blue precipitate which forms a deep blue solution with excess ammonia and
c zinc ions give a white precipitate which forms a colourless solution with excess ammonia.

Vanadium chemistry

- **Vanadium compounds exist in four oxidation states:**

  +5: VO_3^-(aq) which is colourless, but in acid solution reacts to form a **yellow** solution of VO_2^+(aq) ions:
  $$VO_3^- + 2H^+ \rightleftharpoons VO_2^+ + H_2O$$
  +4: VO^{2+}(aq), which is **blue**.
  +3: V^{3+}(aq) which is **green** and is mildly reducing.
  +2: V^{2+}(aq) which is **lavender** and is strongly reducing.

- **Reduction by zinc.** If zinc and dilute hydrochloric acid are added to a solution of sodium vanadate, $NaVO_3$, the colour changes in distinct sequence, as it is steadily reduced from the +5 to the +2 state:

  start: yellow caused by VO^{2+}(aq)ions (+5 state)
  then: green caused by a mixture of yellow VO_2^+(aq) and blue VO^{2+}(aq)
  then: blue caused by VO^{2+}(aq) ions (+4 state)
  then: green caused by V^{3+}(aq) ions (+3 state)
  finally: lavender caused by V^{2+}(aq) ions (+2 state)

- **Redox reactions**

Vanadium compounds can be oxidised or reduced by suitable reagents. For a redox reaction to work, the $E^{\ominus}_{cell}$ value must be positive. A list of $E^{\ominus}$values for the half reactions of vanadium in its various oxidation states and for some oxidising and reducing agents is given below. These values show, for instance, that vanadium(V) will be reduced only to vanadium (IV) by Fe^{2+} ions, whereas Sn^{2+} ions will reduce it first to vanadium (IV) and then to vanadium (III).

$VO_2^+ + 2H^+ + e^- \rightleftharpoons VO^{2+} + H_2O$	$E^{\ominus} = +1.0$ V
$Fe^{3+} + e^- \rightleftharpoons Fe^{2+}$	$E^{\ominus} = +0.77$ V
$\frac{1}{2}Cl_2 + e^- \rightleftharpoons Cl^-$	$E^{\ominus} = +1.36$ V
$VO^{2+} + 2H^+ + e^- \rightleftharpoons V^{3+} + H_2O$	$E^{\ominus} = +0.34$ V
$Sn^{4+} + 2e^- \rightleftharpoons Sn^{2+}$	$E^{\ominus} = +0.15$ V
$Fe^{3+} + e^- \rightleftharpoons Fe^{2+}$	$E^{\ominus} = +0.77$ V
$V^{3+} + e^- \rightleftharpoons V^{2+}$	$E^{\ominus} = -0.26$ V
$Zn^{2+} + 2e^- \rightleftharpoons Zn$	$E^{\ominus} = -0.76$ V
$H^+ + e^- \rightleftharpoons \frac{1}{2}H_2$	$E^{\ominus} = 0.00$ V

+5	VO_2^+	+5
$Fe^{2+} \downarrow$		$\uparrow Cl_2$
+4	VO^{2+}	+4
$Sn^{2+} \downarrow$		$\uparrow Fe^{3+}$
+3	V^{3+}	+3
$Zn/H^+ \downarrow$		$\uparrow H^+$
+2	V^{2+}	+2

$E^{\ominus}_{cell}$ for the reduction of:

VO_2^+ by Fe^{2+} = $+1.0 - (+0.77)$ = $+0.23$ V

VO^{2+} by Sn^{2+} = $+0.34 - (+0.15)$ = $+0.19$ V

V^{3+} by Zn   = $-0.26 - (-0.76)$ = $+0.50$ V

$E^{\ominus}_{cell}$ for the oxidation of:

V^{2+} by H^+ = $0.0 - (-0.26)$ = $+0.26$ V

V^{3+} by Fe^{3+} = $+0.77 - (+0.34)$ = $+0.43$ V

VO^{2+} by Cl_2 = $+1.36 - (+1.0)$ = $+0.36$ V

 Checklist

Before attempting the questions on this topic, check that you:

☐ Can define d block and transition elements.

☐ Can write the electronic structure of the d block elements and their ions.

❑ Can recall the characteristic properties of the transition elements.

❑ Understand the nature of the bonding in complex ions, the shape of the ions and their colour.

❑ Can recall the action of aqueous sodium hydroxide and ammonia on solutions of their aqua ions.

❑ Understand deprotonation and ligand exchange reactions.

❑ Can recall the oxidation states of vanadium, and the colours of its ions.

❑ Know reactions to interconvert the oxidation states of vanadium.

❑ Can recall examples of vanadium, iron and nickel and their compounds as catalysts.

 Testing your knowledge and understanding

Answers
V is [Ar], $3d^3$, $4s^2$. V^{3+} is [Ar], $3d^2$.
$[Cu(H_2O)_6]^{2+}$ turquoise-blue.
$[Cu(NH_3)_4(H_2O)_2]^{2+}$ dark blue
Dative covalent (ligand to ion) and covalent (within the water molecule)
Octahedral
Deprotonation
Ligand exchange
+5, VO_2^+, yellow. +4, VO^{2+}, blue. +3, V^{3+}, green. +2, V^{2+}, lavender.
Iron in the Haber process to manufacture ammonia
Vanadium(V) oxide in the Contact process to manufacture sulphuric acid

 The answers to the numbered questions are on pages 137–138.

For the following set of questions, cover the margin of the page before you answer, then check to see if you are correct.

● Write down the electronic structure of a vanadium atom and a V^{3+} ion.

● State the formula and colour of the aqua complex of copper(II).

● State the formula and colour of the ammine complex of copper(II).

● Name the types of bonding in the aqua complex of iron(II).

● What is the shape of the iron(II) aqua complex ion?

● State the type of reaction that occurs when aqueous sodium hydroxide is added to a solution of the aqua complex of copper(II).

● State the type of reaction that occurs when excess aqueous ammonia is added to a solution of the aqua complex of copper(II).

● What are the oxidation states of vanadium? Give the formula and the colour of the cation for each oxidation state.

● Give an example of the use of iron as an industrial catalyst.

● Give an industrial use of a vanadium compound as a catalyst.

1 Explain why the $[Cu(H_2O)_6]^{2+}$ ion is coloured.

2 Give the equations for the reactions caused by small additions of sodium hydroxide solution, followed by excess to:

   a a solution of the aqua complex of chromium(III)
   b a solution of the aqua complex of iron(II)
   c a solution of the aqua complex of zinc(II).

3 Give the equations for the reactions caused by small additions of ammonia solution, followed by excess to:

   a a solution of the aqua complex of iron(III)
   b a solution of the aqua complex of copper(II).

4 Write the half equations for the following changes in oxidation state:

   a vanadium(V) to vanadium(II)
   b vanadium(V) to vanadium (IV)
   c vanadium(III) to vanadium(IV)

Topic **5.3** Organic chemistry III

 Things to learn and understand

The structure of benzene and reactions of aromatic compounds

● The benzene ring does **not** consist of alternate double and single bonds.

● All the bond lengths are the same and the molecule is planar.

● There is an overlap of p orbitals above and below the plane of the molecule, forming a continuous or **delocalised** π system.

● This gives stability to the benzene structure which is why it undergoes substitution rather than addition reactions.

● Benzene can be represented either by:

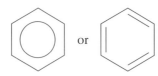

Reactions of benzene

Benzene reacts mainly by **electrophilic** substitution:

● **Nitration**. Benzene reacts with a mixture of concentrated nitric and sulphuric acids to form nitrobenzene:

$$C_6H_6 + HNO_3 \rightarrow C_6H_5NO_2 + H_2O$$

● **Bromination**. Benzene reacts with liquid bromine in the presence of a catalyst of anhydrous iron(III) bromide (made in situ from iron and liquid bromine) or of anhydrous aluminium bromide:

$$C_6H_6 + Br_2(l) \rightarrow C_6H_5Br + HBr$$

● **Friedel–Crafts reaction**. Benzene will react with halogenoalkanes or with acid chlorides in the presence of a catalyst of anhydrous aluminium chloride:

$$C_6H_6 + C_2H_5Cl \rightarrow C_6H_5C_2H_5 + HCl$$
$$\text{ethylbenzene}$$
$$C_6H_6 + CH_3COCl \rightarrow C_6H_5COCH_3 + HCl$$
$$\text{phenylethanone (a ketone)}$$

Reactions of compounds with a carbon-containing side chain

● Compounds such as methyl benzene $C_6H_5CH_3$ and ethyl benzene $C_6H_5C_2H_5$ can be oxidised, and the product will contain a COO^- group on the benzene ring, regardless of the number of carbon atoms in the chain. On acidification benzoic acid is produced:

$$C_6H_5C_2H_5 + 6[O] + OH^- \rightarrow C_6H_5COO^- + CO_2 + 3H_2O$$

conditions: heat under reflux with potassium manganate(VII) and aqueous sodium hydroxide.

Phenol

> Phenol is a weaker acid than carbonic acid, and so it will not liberate CO_2 from sodium hydrogen carbonate, unlike carboxylic acids which will.

Although it forms some hydrogen bonds with water, it is only partially soluble and forms two layers when added to water at room temperature.

● **+ alkali**. Phenol is a weak acid and it forms a colourless **solution** with aqueous sodium hydroxide:

$$C_6H_5OH(l) + OH^-(aq) \rightarrow C_6H_5O^-(aq) + H_2O(l)$$

● **+ bromine water**. The presence of the OH group makes substitution into the benzene ring much easier. No catalyst is required:

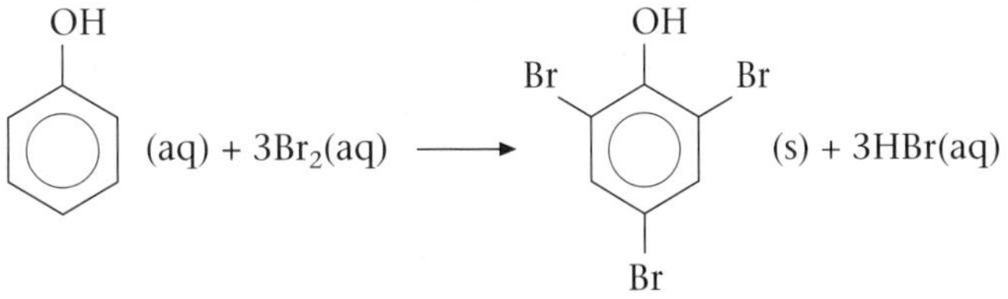

(aq) + 3Br$_2$(aq) ⟶ (s) + 3HBr(aq)

observation: when brown bromine water is added, a white precipitate is rapidly formed.

● **+ acid chlorides**. Phenol reacts as an alcohol, and an ester is formed:

$$C_6H_5OH + CH_3COCl \rightarrow CH_3COOC_6H_5 + HCl$$

Phenylamine

Phenylamine has a NH$_2$ group attached to the benzene ring.

Preparation

Nitrobenzene is reduced by tin and concentrated hydrochloric acid when heated under reflux. Then sodium hydroxide is added to liberate the phenylamine, which is removed by steam distillation.

Reactions

● + acid

$$C_6H_5NH_2 + H^+(aq) \rightarrow C_6H_5NH_3^+(aq)$$

● + nitrous acid

$$C_6H_5NH_2 + 2H^+ + NO_2^- \rightarrow C_6H_5N_2^+ + 2H_2O$$

conditions: add dilute hydrochloric acid to phenylamine and cool the solution to 5 °C. Sodium nitrite solution is then added, keeping the temperature close to 5 °C and above 0 °C.

> The solution of diazonium ions can react with phenol. A yellow precipitate of diazo compound is obtained.
>
> ⟨ ⟩—N＝N—⟨ ⟩—OH

Unit 5

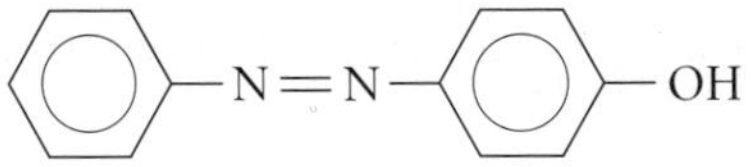

Mechanisms

A curly arrow ($\curvearrowright$) represents the movement of a **pair** of electrons, either from a bond or from a lone pair. A half-headed arrow ($\curvearrowright$) represents the movement of a single electron.

Homolytic, free radical substitution

Reaction between an alkane and chlorine or bromine

● An example is shown of reaction between methane and chlorine.
● There are three stages in this type of reaction:

1 **Initiation**: light energy causes homolytic fission of chlorine:
$$Cl_2 \rightarrow 2Cl\cdot$$

2 **Propagation**: each propagation step involves a radical reacting with a molecule to produce a new radical:
$$CH_4 + Cl\cdot \rightarrow CH_3\cdot + HCl$$
then $$CH_3\cdot + Cl_2 \rightarrow CH_3Cl + Cl\cdot \text{ etc.}$$

3 **Termination**: involves two radicals joining with no radicals being produced, e.g.
$$CH_3\cdot + CH_3\cdot \rightarrow C_2H_6$$

Homolytic, free radical addition

Polymerisation of ethene

The conditions are a very high pressure (1000 atm) and a trace of oxygen. The conditions produce radicals, $R\cdot$, which attack an ethene molecule:
$$R\cdot + H_2C{=}CH_2 \rightarrow R\text{-}CH_2\text{-}CH_2\cdot$$
then $$R\text{-}CH_2\text{-}CH_2\cdot + H_2C{=}CH_2 \rightarrow R\text{-}CH_2\text{-}CH_2\text{-}CH_2\text{-}CH_2\cdot \text{ etc}$$

Heterolytic, electrophilic addition

Reactions of alkenes with halogens or hydrogen halides

The reaction takes place in two steps:

1 The electrophile accepts the π electrons to form a new bond with one of the carbon atoms and at the same time the Br–Br bond breaks.
2 The intermediate carbocation then bonds with a Br⁻ ion.

Heterolytic, electrophilic substitution

Nitration of benzene

1 Sulphuric acid is a stronger acid than nitric and so protonates it:
$$H_2SO_4 + HNO_3 \rightarrow H_2NO_3^+ + HSO_4^-$$
then the cation loses water:
$$H_2NO_3^+ \rightarrow H_2O + NO_2^+$$

Helpful hints

Ensure that the curly arrow **starts** from a bond or an atom with a lone pair of electrons.
Ensure that the arrow points towards an atom either forming a negative ion or a new covalent bond.

Another way of polymerising ethene is by using titanium tetrachloride and aluminium triethyl, but this is not a homolytic mechanism.

Helpful hints

With the addition of a hydrogen halide to an unsymmetrical alkene, the hydrogen atom adds to the carbon which already has more hydrogen atoms directly bonded to it.
This is because the secondary carbocation,

e.g.

is more stable than a primary one.

e.g.

Unit 5

Helpful hints

Similar mechanisms occur for bromination, where the catalyst reacts with bromine to provide the electrophile Br^+, and in the Friedel-Crafts reaction where the catalyst reacts with the halogen compound to provide electrophiles such as CH_3C^+O or $C_2H_5^+$.

2 The NO_2^+ ion is the electrophile and attacks the benzene ring.
3 The HSO_4^- ion pulls off a H^+ and reforms H_2SO_4 (the catalyst).

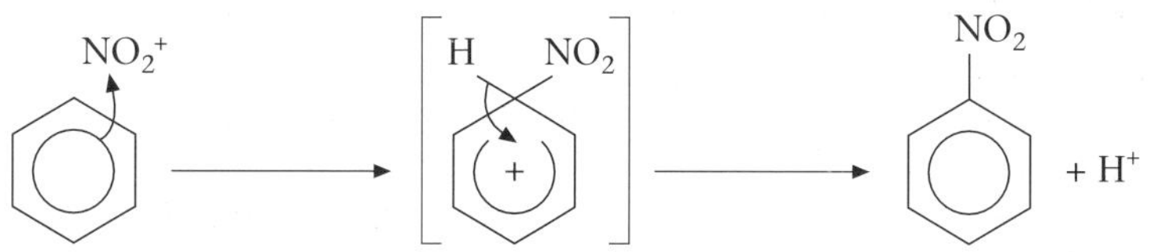

The loss of H^+ results in the reforming of the benzene ring and the gain in stability associated with the ring.

Heterolytic, nucleophilic substitution

Reaction between halogenoalkanes and hydroxide or cyanide ions

● There are two mechanisms depending on the type of halogenoalkane.
● With primary ($1°$) halogenoalkanes, an S_N2 mechanism is dominant.
● Reaction proceeds via a transition state when the lone pair of electrons on the nucleophile attacks the halogenoalkane.

Helpful hints

In the transition state, the new O–C bond forms as the C–Br bond breaks. The transition state has a charge of −1.

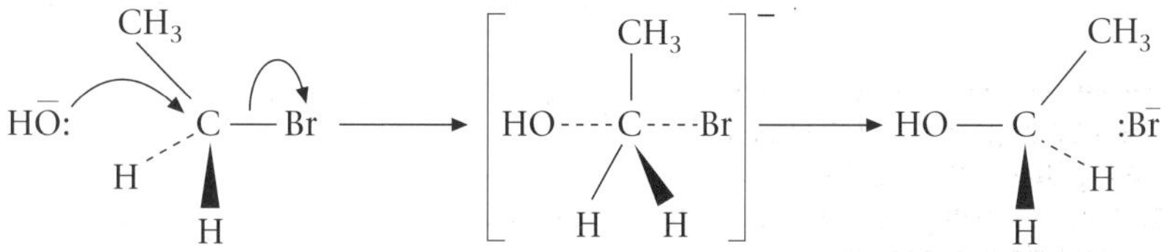

transition state

● With tertiary ($3°$) halogenoalkanes the S_N1 mechanism is dominant. This happens in two steps:
1 The halogenoalkane ionises in the relatively slow, and hence rate determining step, to form an **intermediate** carbocation.

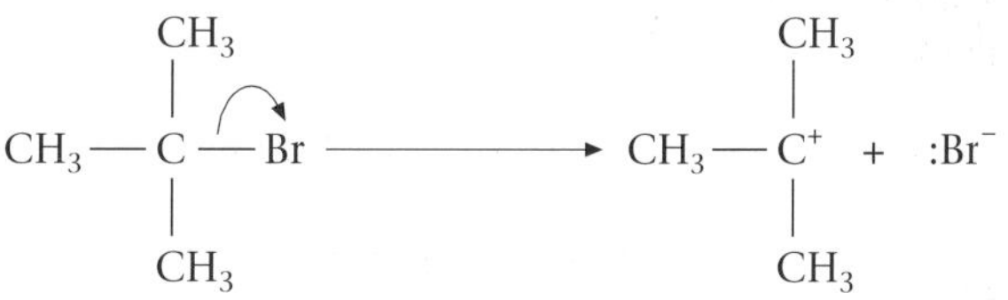

Helpful hints

Any optical activity is maintained in a S_N2 reaction.
If the mechanism is S_N1, a reactant which is an optical isomer will produce a racemic mixture, as the carbocation can be attacked from either side.

2 This then rapidly forms a bond with the nucleophile, OH^-.

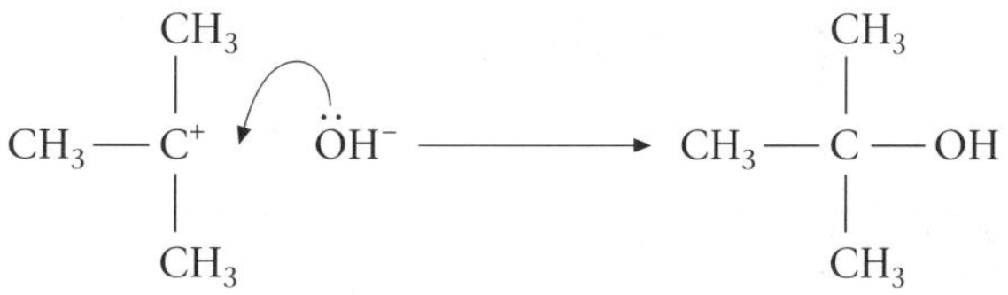

Heterolytic, nucleophilic addition

Reaction between a carbonyl compound and hydrogen cyanide

● Although the reaction is the addition of HCN, the first step is the nucleophilic attack by the CN^- ion on the $\delta+$ carbon atom in the carbonyl group.

Helpful hints

The CN^- ion is a catalyst, and the conditions must be such that there is a significant amount of both HCN molecules and CN^- ions present, i.e. a mixture of HCN and KCN.

- The negatively charged oxygen in the anion then removes an H^+ from an HCN molecule. This produces another CN^- ion, which reacts with another carbonyl group and so on.

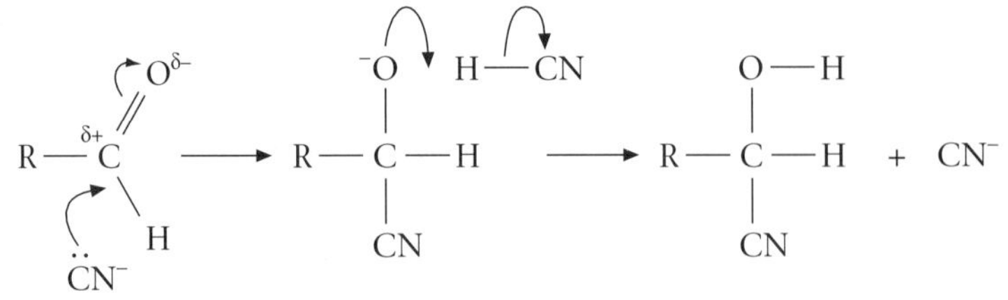

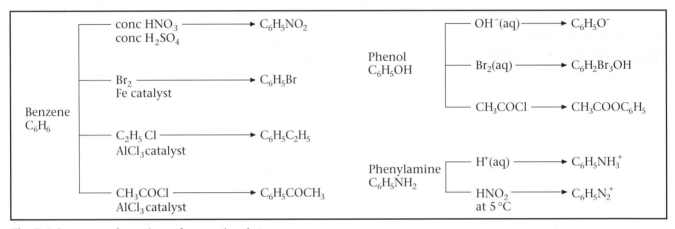

- The conditions for this reaction are either HCN and a trace of base or KCN and a small amount of dilute sulphuric acid.

Reactions of aromatic substances (Figure 5.4) and mechanisms (Figure 5.5) are summarised schematically.

Benzene C_6H_6
- conc HNO_3 / conc H_2SO_4 → $C_6H_5NO_2$
- Br_2 / Fe catalyst → C_6H_5Br
- C_2H_5Cl / $AlCl_3$ catalyst → $C_6H_5C_2H_5$
- CH_3COCl / $AlCl_3$ catalyst → $C_6H_5COCH_3$

Phenol C_6H_5OH
- $OH^-(aq)$ → $C_6H_5O^-$
- $Br_2(aq)$ → $C_6H_2Br_3OH$
- CH_3COCl → $CH_3COOC_6H_5$

Phenylamine $C_6H_5NH_2$
- $H^+(aq)$ → $C_6H_5NH_3^+$
- HNO_2 at 5 °C → $C_6H_5N_2^+$

Fig 5.4 *Summary of reactions of aromatic substances*

	Substitution				**Addition**		
Free radical:	alkanes	+	Cl_2	Free radical:	polymerisation of alkenes		
Nucleophilic:	halogenoalkanes	+	$OH^-(aq)$	Nucleophilic:	carbonyl compounds	+	HCN
		+	CN^-				
Electrophilic:	benzene	+	HNO_3	Electrophilic:	alkenes	+	Cl_2/Br_2
		+	$Br_2(l)$			+	HCl/HBr/HI
		+	RCl				
		+	RCOCl				

Fig 5.5 *Summary of mechanisms*

 *Checklist*

Before attempting the questions on this topic, check that you:

☐ Understand the structure of benzene and why it reacts by substitution rather than by addition.

☐ Can recall the reactions of benzene with nitric acid, bromine and halogen compounds.

☐ Can recall the product of oxidising side chains.

☐ Can recall the reactions of phenol with alkali, bromine and acid chlorides.

☐ Can recall the preparation of phenylamine and its reaction with nitrous acid and the coupling of the product with phenol.

☐ Understand the mechanisms of free radical substitution and addition.

☐ Understand the mechanisms of electrophilic substitution and addition.

☐ Understand the mechanisms of nucleophilic substitution and addition.

 Testing your knowledge and understanding

Answers
a $C_6H_5COO^-$
b $C_6H_5O^-$ Na^+
c $C_2H_5COOC_6H_5$

The brown bromine water would go colourless and a white precipitate would form.

Heat under reflux with tin and concentrated hydrochloric acid, cool and then add aqueous sodium hydroxide.

 The answers to the numbered questions are on page 138.

For the following set of questions, cover the margin before you answer, then check to see if you are correct.

● State the structural formulae of the organic product obtained by the reaction of:

  a propylbenzene with alkaline potassium manganate(VII)

  b phenol with sodium hydroxide solution

  c phenol with propanoyl chloride.

● What would you observe if bromine water were added to aqueous phenol?

● State the conditions for the conversion of nitrobenzene to phenylamine.

1 a Describe the mechanism for the reaction of bromine with benzene.

  b Why does benzene undergo a substitution reaction with bromine rather than an addition reaction?

2 In the reaction between propanone and hydrogen cyanide no reaction occurs unless a small amount of a base such as sodium hydroxide is added. Explain these observations.

3 When benzenediazonium chloride is prepared, the reaction is carried out at about 5 °C.

  a Why are these conditions chosen?

  b What is the formula of the product obtained by the reaction of the benzene diazonium chloride solution with phenol in alkaline solution?

Unit 5

Topic 5.4 Chemical kinetics II

Introduction

You must also revise Topic 2.3 in Chapter 2, page 44 and especially how the changes in the Maxwell–Boltzmann distribution of energy with increase in temperature affect the rate of reaction.

Things to learn

> The units of *k* are:
> - for a zero-order reaction: $mol\ dm^{-3}\ s^{-1}$
> - for a first-order reaction: s^{-1}
> - for a second-order reaction: $mol^{-1}\ dm^3\ s^{-1}$

❏ **The rate equation** for the reaction:

$$x\mathrm{M} + y\mathrm{N} \rightarrow \text{products, is:}$$

Rate of reaction = $k\,[\mathrm{M}]^a\,[\mathrm{N}]^b$ where *a* and *b* are integers and are **experimentally** determined.

❏ **The rate constant, *k*,** is the constant of proportionality in the rate equation:

- Its value depends on the activation energy of the reaction and the temperature.
- Reactions with a large activation energy will have small values of *k*.

❏ **The order with respect to one substance** is the power to which the concentration of that substance is raised in the rate equation. In the example above the partial order of the chemical M is *a*.

❏ **The order of reaction** is the sum of the partial orders. In the example above, the order of the reaction is *a* + *b*.

❏ **The activation energy, E_a,** is the total kinetic energy that the molecules must have on collision in order for them to be able to react.

> The half-life of a first-order reaction is constant.

❏ **Half-life, $t_{1/2}$,** is the time taken for the concentration to fall from any selected value to half that value.

Things to understand

Determination of rate equation from initial rates

Consider the reaction:

$$\mathrm{A} + \mathrm{B} + \mathrm{C} \rightarrow \text{products}$$

The initial rates of reaction with different concentrations of A, B and C are found, and by taking the experiments two at a time, the order with respect to each substance can be found.

Look for two experiments where the concentration of only **one** substance varies. If doubling that concentration causes the rate to double, the order with respect to that substance is 1.

Experiment	[A]	[B]	[C]	Relative rate
1	1	1	1	1
2	2	1	1	2
3	1	2	1	4
4	1	2	2	4

- From experiments 1 and 2: [A] doubles and rate doubles. Therefore order with respect to A = 1
- From experiments 1 and 3: [B] doubles and rate × 4. Therefore order with respect to B = 2
- From experiments 3 and 4: [C] doubles and rate unaltered. Therefore order with respect to C = 0
- Rate of reaction = k $[A]^1$ $[B]^2$ $[C]^0$, and the reaction is 3rd order.

Experimental techniques for following a reaction

1 Titration method. Mix the chemicals and start the clock. At intervals withdraw a sample, add it to iced water to slow the reaction. Then titrate one of the substances in the reaction. This method can be used if an acid, an alkali or iodine is a reactant or a product.

2 Colorimetric method. This can be used when either a reactant or a product is coloured (iodine or potassium manganate(VII) are examples). The colorimeter must first be calibrated using solutions of the coloured substance of known concentrations. Then the reactants are mixed and the clock started. The intensity of the colour is measured as a function of time. The concentration of the coloured substance is proportional to the amount of light absorbed.

3 Gas volume method. If a gas is produced in the reaction, its volume can be measured at intervals of time. The gas can be collected in a horizontal gas syringe or by bubbling it into an inverted measuring cylinder filled with water.

Graphs of results

If the concentration of a reactant is plotted against time, three shapes of graph are likely depending on the order of the reaction.

Mechanisms

A suggested mechanism must be consistent with the order of reaction. The partial order of any species which occurs in the mechanism **after** the rate determining step will be zero. The rate-determining step is the slowest step, e.g.

$$NO_2(g) + CO(g) \rightarrow NO(g) + CO_2(g)$$ has the rate expression:

$$\text{Rate} = k \, [NO_2]^2 \, [CO]^0$$

A suggested mechanism is:
- Step 1 (slow):

$$NO_2 + NO_2 \rightarrow NO_3 + NO$$

- Step 2 (fast):

$$NO_3 + CO \rightarrow NO_2 + CO_2$$

As the CO enters the mechanism **after** the rate determining step, the mechanism is consistent with the fact that the reaction is zero order with respect to CO.

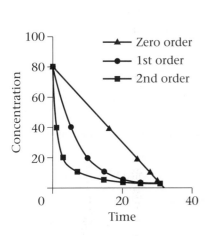

Unit 5

Energy profile diagrams

The first diagram is for a reaction that takes place in a single step, and the second diagram is for this reaction with a catalyst (see Figure 5.6).

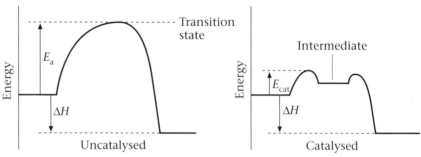

Fig 5.6 *Energy profile diagrams*

An example of a reaction with a transition state is the S_N2 reaction of bromoethane with hydroxide ions.

An example of a catalysed reaction forming an intermediate is Fe^{2+} ions as a catalyst in:

$$S_2O_8^{2-}(aq) + 2I^-(aq) \rightarrow 2SO_4^{2-}(aq) + I_2(aq)$$

In this mechanism the catalyst first reduces the $S_2O_8^{2-}$ ions:

$$2Fe^{2+}(aq) + S_2O_8^{2-}(aq) \rightarrow 2Fe^{3+}(aq) + 2SO_4^{2-}(aq)$$

then the Fe^{3+} ions are reduced back to Fe^{2+} ions:

$$2Fe^{3+}(aq) + 2I^-(aq) \rightarrow 2Fe^{2+}(aq) + I_2(aq)$$

 Checklist

Before attempting the questions on this topic, check that you:

❑ Can define rate constant, order of reaction and half-life.

❑ Can deduce rate equations from initial rate data.

❑ Understand the concept of activation energy and its relation to the rate constant.

❑ Understand that information about mechanisms can be deduced from the partial orders of the reactants.

❑ Can recall the energy profiles of reactions with and without catalysts.

❑ Can suggest suitable methods for following reactions.

❑ Can deduce the order of a reaction from concentration/time graphs.

Testing your knowledge and understanding

For the following set of questions, cover the margin before you answer, then check to see if you are correct.

● The following results were obtained from a study of the reaction:

$$NO_2(g) + CO(g) \rightarrow NO(g) + CO_2(g)$$

Experiment	$[NO_2]$/mol dm^{-3}	$[CO]$/mol dm^{-3}	Relative rate
1	0.02	0.02	1
2	0.04	0.02	4
3	0.02	0.04	1

a What is the order with respect to NO_2?
b What is the order with respect to CO?
c What is the order of reaction?
d State the rate expression.

● Reaction A has a high value of E_a, and reaction B has a lower E_a value.
a Which reaction has the larger rate constant?
b Which is the faster reaction?

● A reaction takes place in 3 steps. The E_a/kJ mol^{-1} for each step are:
   step 1: 32,   step 2: 51,   step 3: 20
Which step determines the rate?

● The decomposition of N_2O_5 is first order. At 200 °C the reaction has a half-life of 25 minutes. Calculate how long it will take for the concentration of N_2O_5 to fall to 6.25% of its original value.

1 The rate of the second order reaction:
$$2HI(g) \rightarrow H_2(g) + I_2(g)$$
is 2.0×10^{-4} mol dm^{-3} s^{-1} when [HI] = 0.050 mol dm^{-3} at 785 K. Calculate the value of the rate constant, giving its units.

2 Describe how you would follow, at 60 °C, the rate of the reaction:
$$CH_3COOH(aq) + CH_3OH(aq) \rightarrow CH_3COOCH_3(l) + H_2O(l).$$

3 The decomposition of 3-oxobutanoic acid, CH_3COCH_2COOH, was studied:
$$CH_3COCH_2COOH \rightarrow CH_3COCH_3 + CO_2$$
The results, at 40 °C, are tabulated below.

Time/min	[3-oxobutanoic acid]/mol dm^{-3}
0	1.6
26	0.8
52	0.4
78	0.2

Deduce the order of the reaction.

Answers

a From experiments 1 and 2, order with respect to NO_2 = 2
b From experiments 1 and 3, order with respect to CO = 0
c Order of reaction is 2 + 0 = 2
d Rate = k $[NO_2]^2$

a Reaction B
b Reaction B

Step 2 as it has the largest E_{act} value

100 to 6.25 = 4 half-lives, therefore time = 4 x 25 = 100 minutes

 The answers to the numbered questions are on page 138.

Unit 5

Topic 5.5 Organic chemistry IV

Introduction

- Questions on this topic will form part of the synoptic assessment in Unit Test 5.
- This Topic brings together all the organic chemistry that has been covered in earlier topics, so you will need to revise Topics 2.2, 4.5 and 5.3.

Things to learn and understand

Organic analysis

You should know the tests for:

- **Alkenes**
  Test: Add bromine in hexane.
  Observation: The brown bromine becomes colourless.

- **Halogenoalkanes**
  Test: Heat under reflux with sodium hydroxide solution, then acidify with dilute nitric acid. Add silver nitrate solution.
  Observation: Chlorides give a white precipitate that is soluble in dilute ammonia.
  Bromides give a cream precipitate that is insoluble in dilute ammonia but dissolves in concentrated ammonia.
  Iodides give a yellow precipitate that is insoluble in concentrated ammonia.

- **OH group** (in alcohols and acids)
  Test: To the **dry** compound add phosphorus pentachloride.
  Observation: Steamy fumes of hydrogen chloride produced.

- **Alcohols**
  Test: Warm with dilute sulphuric acid and potassium dichromate(VI) solution.
  Observation: **Primary** (1°) and **secondary** (2°)alcohols reduce the orange dichromate(VI) ions to green Cr^{3+} ions.
  Tertiary (3°) alcohols do not affect the colour as they are not oxidised.
  To distinguish between 1° and 2° repeat the experiment, but distil the product into ammoniacal silver nitrate solution:
  1° alcohols are oxidised to aldehydes, which give a silver mirror.
  2° alcohols are oxidised to ketones, which do not react.

- **Carbonyl, C=O group** (aldehyde or ketone)
  Test: Add a solution of 2,4-dinitrophenylhydrazine.
  Observation: A red or orange precipitate is seen.

To distinguish between aldehydes and ketones:

Test: Warm with ammoniacal silver nitrate solution.

Observation: Aldehydes give a silver mirror, ketones have no reaction.

● **Iodoform reaction**

Test: Gently warm with a solution of sodium hydroxide and iodine.

Observation: A pale yellow precipitate of CHI_3

This test works with carbonyl compounds containing the $CH_3C=O$ and alcohols containing the $CH_3CH(OH)$ groups.

● **Carboxylic acids, COOH group**

Test: Add to a solution of sodium hydrogen carbonate.

Observation: Bubbles of gas, which turn lime water milky.

Interpretation of data

You should be able to deduce structural formulae of organic molecules given data obtained from:

● **Chemical methods**

Carry out tests as above to determine the functional groups in the molecule. Measure the melting point of the substance. Make a solid derivative (such as the 2,4-DNP derivative from carbonyl compounds), purify it, and measure its melting temperature. Check the melting temperature values with a data bank to identify the substance.

● **Mass spectra** (see Figure 5.7)
  ● Observe the fragments obtained and look for the value of the molecular ion $(M)^+$, if any.
  ● See if there is a peak at $(M-15)^+$. If so the substance probably had a CH_3 group which is lost forming the $(M-15)^+$ peak.
  ● If there is a peak at $(M-28)^+$, the substance probably had a CO group. Other fragments will help to identify the structure.

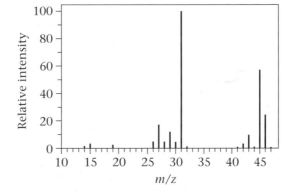

> The solid must be purified by recrystallisation in order to have a sharp and accurate melting temperature.

> *Helpful hints*
>
> When identifying species that cause lines in the spectrum, always give a structural formula for the ion and its + charge.

Fig 5.7 *Mass spectrum of ethanol*

> The peaks given by O-H and N-H are usually broad owing to hydrogen bonding.

● **Infrared spectra** (see Figure 5.8)
  ● These are used to identify functional groups in the substance, and also to compare the spectrum of the unknown with a data bank of spectra ('finger printing').
  ● Peaks to look for are at:

1650 to 1750 cm^{-1}	given by C=O
2500 to 3500 cm^{-1}	given by O–H
3300 to 3500 cm^{-1}	given by N–H
1000 to 1300 cm^{-1}	given by C–O

Unit 5

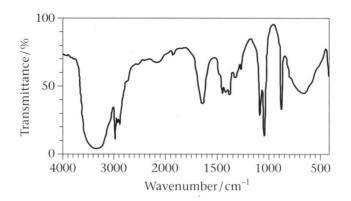

Fig 5.8 *Infrared spectrum of ethanol*

The chemical shift is the difference between the absorption frequencies of the hydrogen nuclei in the compound and those in the reference compound.

NMR spectra

- The phenomenon of nuclear magnetic resonance occurs when nuclei such as ^1H are placed in a strong magnetic field and then absorb applied radio frequency radiation.
- The nuclei of hydrogen atoms in different chemical environments within a molecule will show up separately in a NMR spectrum. The values of their chemical shift, δ, are different.
- The hydrogen nuclei in a CH_3 group will have a different chemical shift from those in a CH_2 or in an OH group. The value of δ of the peak due to the hydrogen in OH (or in NH) depends upon the solvent.
- In **low resolution** NMR, each group will show as a single peak, and the area under the peak is proportional to the number of hydrogen atoms in the same environment. Thus ethanol, CH_3CH_2OH will have three peaks of relative intensities 3:2:1 and methyl propane $CH_3CH(CH_3)CH_3$, will have two peaks with relative intensities of 9:1.
- In **high resolution** NMR spin coupling is observed. This is caused by the interference of the magnetic fields of neighbouring hydrogen nuclei. If an adjacent **carbon** atom has hydrogen atoms bonded to it, they will cause the peaks to split as follows:

  1 neighbouring H atom   peak splits into 2 lines (a doublet)
  2 neighbouring H atoms   peak splits into 3 lines (a triplet)
  n neighbouring H atoms   peak splits into $(n + 1)$ lines.

  Thus ethanol gives three peaks (see Figure 5.9):

  1 peak due to the OH hydrogen, which is a single line
      (as it is hydrogen bonded)
  1 peak due to the CH_2 hydrogens, which is split into four lines
      by the three H atoms on the neighbouring CH_3 group.
  1 peak due to the CH_3 hydrogens, which is split into three lines
      by the two H atoms on the neighbouring CH_2 group.

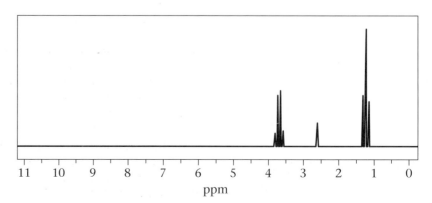

Fig 5.9 *High resolution NMR spectrum of ethanol*

● **Visible and UV spectra**
  ● π bonds will absorb light energy as an electron is promoted from a bonding to an anti-bonding π orbital.
  ● The energy gap is in the near UV or visible range, and the group causing this is called a chromophore.
  ● If the π bonded electrons are linked to a conjugated system (alternate double and single carbon–carbon bonds), the absorption will shift to a lower frequency, and the substance may be coloured. The colour observed will be the complementary colour of that absorbed.

Organic synthesis

Reagents and conditions for the preparation of substance B from A

If the number of carbon atoms in the chain is:
  ● **increased** by one, consider:
    a halogenoalkane with KCN
    b carbonyl with HCN
    c the Grignard reagent CH_3MgBr
    d a Grignard reagent added to CO_2 or methanal
  ● **increased** by **more** than one, consider:
    a a Grignard reagent with more than one carbon atom
    b Friedel–Crafts reaction for aromatic substances
  ● **decreased** by one, consider:
    a the Hofmann degradation
    b the iodoform reaction.

Practical techniques

● **Recrystallisation**. Dissolve the solid in a **minimum** of **hot** solvent. Filter the hot solution through a **preheated** funnel using **fluted** filter paper. Allow to **cool**. Filter under **reduced pressure** (Buchner funnel), **wash** with a little cold solvent and allow to **dry**. The solid should have a sharp melting temperature.
● **Heating under reflux** is necessary when either the reactant has a low boiling temperature or the reaction is slow at room temperature.
● **Safety precaution**. You must use a fume cupboard when a reactant or product is toxic, irritant or is carcinogenic.
● **Fractional distillation**. This is used to separate a mixture of two liquids. These mixtures can usually be separated into pure samples, but if the boiling points are too close, a good separation will be difficult.

If a liquid of composition X is heated, it will boil at T_1 °C to give a vapour of composition Y (see Figure 5.10). If this is condensed and reboiled, it will boil at a temperature of T_2 °C giving a vapour of composition Z. Eventually pure B will distill off the top and pure A will be left in the flask.

Applied organic chemistry

● **Targeting pharmaceutical compounds**. Those which are ionic or have several groups that can form hydrogen bonds with water, will tend to be retained in aqueous (non-fatty) tissue. Compounds with long hydrocarbon chains and with few hydrogen bonding groups will be retained in fatty tissue. The latter will be stored in the body, whereas water-soluble compounds will be excreted.
● **Nitrogenous fertilisers** are of three types:
  1 quick release, such as those containing NO_3^- and NH_4^+ ions
  2 slow release, such as urea, NH_2CONH_2
  3 natural, such as slurry, compost and manure.
The first two are water-soluble and can be leached out. The last is bulky and contains little nitrogen.

Helpful hints

1 A common error is to think that KCN or HCN will react with alcohols to form RCN. You must first convert ROH to R–halogen.
2 Many questions on organic synthesis require, somewhere in the sequence, the conversion of the starting substance to a carbonyl compound or a halogenoalkane.

Helpful hints

Recrystallisation is only suitable for purifying solids.

Helpful hints

When a specific safety precaution is asked for, do not give the use of safety glasses, lab coats etc. as examples. The answer is probably 'no naked flames (ether solvent)' or 'carry out the experiment in a fume cupboard', in which case the specific hazard such as toxicity must be stated.

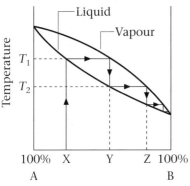

Fig 5.10 *Boiling temperature/composition diagram*

● **Esters, oils and fats.** Simple esters are used in food flavourings, perfumes and as solvents for glues, varnishes and spray paints. Animal fats are esters of propan-1,2,3-triol and **saturated** acids such as stearic acid $C_{17}H_{35}COOH$. When hydrolysed by boiling with aqueous sodium hydroxide, they produce soap which is the sodium salt of the organic acid.

Vegetable oils are liquids which are esters of propan-1,2,3-triol and **unsaturated** acids.

Margarine is made from polyunsaturated vegetable oils by partial hydrogenation (addition of hydrogen) so that only a few double bonds remain.

> Polymers usually soften over a range of temperatures rather than have a sharp melting temperature. The main reason for this is that the polymer is a mixture of molecules of different chain lengths.

● **Polymers** are of two distinct chemical types:

1   **Addition**. The monomers contain one or more C=C groups. Polymerisation is the addition caused by breaking the π bonds, and the polymer and the monomer have the same empirical formula. Examples are poly(ethene), poly(tetrafluoroethene) (PTFE) and poly(propene).

2   **Condensation**. Both the monomers have **two** functional groups, one at each end. Polymerisation involves the loss of a simple molecule (usually H_2O or HCl) as each link forms. Examples are nylon, a polyamide (see Figure 5.11), and terylene, a polyester.

| a diacid chloride | a diamine | a polyamide |

Fig 5.11 *A polyamide*

Synthetic polymers are not easily biodegraded and can cause an environmental problem of disposal. One answer is for them to be recycled.

 Checklist

Before attempting the questions on this topic, check that you:

☐ Can recall the tests for C=C, C–Hal, COH, C=O, CHO and COOH groups.

☐ Can deduce the functional groups present from results of chemical tests.

☐ Can interpret mass, IR, NMR and UV spectra.

☐ Can deduce pathways for the synthesis of organic molecules.

☐ Appreciate how the structure of a pharmaceutical affects its solubility.

☐ Can distinguish between types of polymer and understand polymerisation.

 Testing your knowledge and understanding

> The answers to the numbered questions are on pages 138–139.

1   How would you distinguish between:

   a   pentan-3-one and pentanal

    b   2-bromopropane and 2-chloropropane

    c   methylpropan-2-ol, methylpropan-1-ol and butan-2-ol?

2 The mass spectrum of a substance X of molecular formula C_8H_8O had peaks at mass/charge values of 120, 105, and 77. Identify the peaks and suggest a structural formula for X.

3 The IR spectrum of aspirin showed peaks at 3200, 1720 and 1150 cm^{-1}. Use the data on page 109 to identify the groups which cause these peaks.

4 Low-resolution NMR spectra of two aromatic isomers with molecular formula C_8H_{10} were obtained. Isomer Y had three peaks of relative areas 5:2:3, and isomer Z had two peaks of relative areas 6:4. Suggest structural formulae for Y and Z.

Practice Test: Unit 5

Time allowed 1hr 30 min

All questions are taken from parts of previous Edexcel Advanced GCE papers

The answers are on pages 139–141.

1 a The kinetics of the hydrolysis of the halogenoalkane (where R is an alkyl group) RCH_2Cl with aqueous sodium hydroxide was studied at 50 °C. The following results were obtained:

Experiment	[RCH_2Cl]	[OH^-]	Initial rate/mol dm^{-3} s^{-1}
1	0.050	0.10	4.0×10^{-4}
2	0.15	0.10	1.2×10^{-3}
3	0.10	0.20	1.6×10^{-3}

  i Deduce the order of reaction with respect to the halogenoalkane, RCH_2Cl, and with respect to the hydroxide ion, OH^-, giving reasons for your answers. **[4]**
  ii Hence write the rate equation for the reaction. **[1]**
  iii Calculate the value of the rate constant with its units for this reaction. **[2]**
  iv Using your answer to part **ii**, write the mechanism for this reaction. **[3]**
  (Total 10 marks)
  [January 2001CH6 question 1]

2 Benzene, C_6H_6, reacts with ethanoyl chloride, CH_3COCl, by an electrophilic substitution reaction in the presence of aluminium chloride as catalyst.
  a Identify the electrophile involved in this reaction and write the equation to show its formation. **[2]**
  b Draw the mechanism for the electrophilic substitution of benzene by ethanoyl chloride. **[3]**
  c Suggest a reaction scheme, stating reagents and conditions, to convert the product of the reaction in
    b into

$$C_6H_5\overset{\overset{\displaystyle OH}{|}}{\underset{\underset{\displaystyle CH_3}{|}}{C}}\!-\!COOH$$

  [5]
  d Ethanoyl chloride can be used to prepare esters such as 3-methylbutyl ethanoate, $(CH_3)_2CHCH_2CH_2OOCCH_3$, which is a bee alarm pheromone that signals danger to a honey bee. If this compound is warmed with aqueous sodium hydroxide, a slow reaction takes place to produce sodium ethanoate and 3-methylbutan-1-ol. The reaction is first order with respect to **both** 3-methylbutyl ethanoate and the aqueous hydroxide ion. Explain the term *first order* and give experimental details showing how this information could be obtained. **[8]**
  (Total 18 marks)
  [June 2001 CH6 questions 2 & 7]

3 Study the reaction scheme below, and then answer the questions that follow.

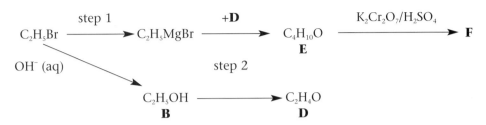

a Give the reagents and conditions necessary for step 1 **[2]**
b Compound **D** reacts with ammoniacal silver nitrate solution to give a silver mirror.
   i Give suitable reagents for the conversion of **B** to **D**. **[2]**
   ii What precautions should be taken to ensure that **B** was converted mainly to **D** and did not react further? **[1]**
   iii Give the name or the structural formula of the organic compound formed when **D** reacts with ammoniacal silver nitrate solution. **[1]**
c Give the structural formula of compound **E** ($C_4H_{10}O$). **[2]**
d Compound **F** reacts with 2,4-dinitrophenylhydrazine solution but has no effect upon ammoniacal silver nitrate solution. **F** will also undergo the iodoform reaction.
   i Explain the significance of the results of the test with ammoniacal silver nitrate and 2,4-dinitrophenyl hydrazine concerning the functional group in **F**. **[2]**
   ii What structural feature in **F** is identified by the iodoform reaction? Give the formula of the products of the reaction. **[3]**
   iii Give the structural formula of **F**. **[2]**
e The mass spectrum of **F** showed major peaks at *m/e* values of 72, 43 and 29. Identify the species responsible for these peaks. **[3]**

(Total 18 marks)

[January 2002 CH4 question 1]

4 a Describe how you would measure the standard electrode (reduction) potential of the $Fe^{3+}(aq)/Fe^{2+}(aq)$ system. **[5]**
b A standard $Fe^{3+}(aq)/Fe^{2+}(aq)$ electrode is connected to a standard gold electrode, $Au^{3+}(aq)/Au(s)$, at 25 °C. The gold electrode acts as the cathode and the electrons flow, in the external circuit, from the $Fe^{3+}(aq)/Fe^{2+}(aq)$ electrode to the gold electrode.
   i Write the half equations for the reactions that take place at each electrode when an electric current is drawn from the cell. **[2]**
   ii Hence write the overall ionic equation for the reaction that takes place. **[1]**
   iii The potential of this cell is +0.73 V and the standard electrode potential for a $Fe^{3+}(aq)/Fe^{2+}(aq)$ electrode is +0.77 V. Calculate the standard electrode potential of the gold electrode. **[2]**

(Total 10 marks)

[January 2001 CH3 question 3]

Unit 5

5 Iron and chromium are very important industrial metals.
  a Complete the boxes below to show the electronic configurations of:

                                    3d   4s

  Fe: [Ar]   ☐   ☐   ☐   ☐   ☐   ☐

  Cr: [Ar]   ☐   ☐   ☐   ☐   ☐   ☐

  Cr^{3+}:   [Ar]   ☐   ☐   ☐   ☐   ☐   ☐   **[3]**

  b Chromite, $FeCrO_4$ or $FeO.CrO_3$, is a mixed oxide of iron and chromium that is used in the manufacture
  of stainless steel. It can be reduced by heating with carbon to produce iron, chromium and carbon
  monoxide.
  Write an equation to represent the reduction of chromite.   **[2]**
  c The mixture of metals that results from the reduction in **b** was reacted with excess dilute sulphuric
  acid to give a solution of iron(II) sulphate and chromium(III) sulphate. Sodium hydroxide solution
  was added to the mixture until in excess. The mixture was then filtered to give a precipitate **Q**,
  containing all the iron, and a filtrate **R**, containing all the chromium.
  i   What is the formula of the iron-containing ion in the aqueous iron(II) sulphate solution?   **[1]**
  ii Write an ionic equation for the reaction between sodium hydroxide solution and this ion. State
        what type of reaction is taking place.   **[3]**
  iii State the colour of the precipitate **Q**.   **[1]**
  iv Give the formula of the chromium-containing ion in **R**.   **[2]**
  v   State what you would see if solution **R** was slowly acidified with dilute sulphuric acid until there
        was no further change.   **[2]**

Electrode reaction	E°/V	Electrode reaction	E°/V
		$\frac{1}{2}F_2 + e^- \rightleftharpoons F^-$	+2.87
$Cr^{3+} + e^- \rightleftharpoons Cr^{2+}$	–0.41	$\frac{1}{2}Cl_2 + e^- \rightleftharpoons Cl^-$	+1.36
$\frac{1}{2}Cr_2O_7^{2-} + 7H^+ + 3e^- \rightleftharpoons Cr^{3+} + 3\frac{1}{2}H_2O$	+1.33	$\frac{1}{2}Br_2 + e^- \rightleftharpoons Br^-$	+1.07
		$\frac{1}{2}I_2 + e^- \rightleftharpoons I^-$	+0.54

  d Use the standard electrode (reduction) potentials of the chromium ions and the halogens shown below
  to answer the questions that follow.
  i   Define the term *standard electrode potential*.   **[2]**
  ii List those halogens which will oxidise chromium(II) to chromium(III).   **[1]**
  iii List those halogens which will oxidise chromium(II) to chromium(III) but **not** to chromium(VI).**[1]**
  iv Chromium(II) in aqueous solution is sky blue whereas aqueous chromium(III) solution is dark
        green. Describe how you would show that your prediction in **d** part **ii** actually worked in practice.
        [2]

(Total 19 marks)
[January 2002 CH1 question 3 & June 2001 CH3 question 5]

6 Advanced laboratory chemistry

Laboratory chemistry II

- The specification (syllabus) for the tests on this module includes all the AS and A2 material.
- The Unit Test 6B is a synoptic paper, taken by all candidates, and will also assess the candidate's quality of written communication.

 ## *Unit Test 6A: Assessment of practical skills II*

❏ This is either internally assessed or a practical exam.

❏ Notes and books may be used in the tests.

❏ The practical exam will be broadly qualitative in its approach.

❏ You should be able to:

   i observe and interpret details of the chemistry of the elements and compounds listed in Units 4 and 5.

   ii recognize the results of experiments to identify functional groups in organic compounds.

  iii carry out the techniques described in Topic 5.5 and those used in volumetric analysis, kinetics and equilibria.

  iv present and interpret quantitative and qualitative results.

   v devise and plan simple experiments based on the chemistry and techniques as above.

 ## *Unit Test 6B*

All candidates will do this paper

The Unit Test question (exam) paper

- **Section A** will consist of a compulsory question and will assess a candidate's ability to interpret data from laboratory situations.
- **Section B** will consist of three questions and the candidate must answer two of the three questions.
- These questions will require candidates to make connections between different areas of chemistry, for example by applying knowledge and understanding of principles to situations in contexts new to them.

- Questions will be set on any of the topics in the AS and A2 specification.

- The questions will require much less factual recall than those in earlier unit tests. Much more emphasis will be placed upon application of knowledge.

- Questions will test a candidate's ability to analyse information from several different areas of the specification.

- The words 'suggest' and 'deduce' will occur more often in questions than 'state' or 'recall'.

Tackling the paper

- Spend some time looking through Section B and decide which questions you are going to attempt. Read all of the question before rejecting it. You may be able to answer all but the first part and so still score good marks. If you are getting nowhere in a question, abandon it and try another, but **do not cross out what you have written**, because you might score more for it than for the other question. The examiner will count your better mark.

- Do not be put off by unusual compounds or situations. In these questions you are not expected to **know** the answer, but to be able to work it out, using your knowledge and understanding of similar compounds or situations.

- Synoptic questions will contain material from several topics. This is done by using the links that exist between different branches of chemistry.

- Each question should have a thread or link connecting the different parts. Identification of this thread will help you to focus on the relevant chemistry. So do not treat each part of a question in isolation from the other parts of it.

- For example in the questions in the Unit Test 6B on pages 119 and 120.

  - Question 2 is based on organic and inorganic nitrogen compounds, with questions on fertilisers, pK_a, polymers, amine preparation, and K_p for ammonium nitrate linked to thermodynamic and kinetic stability.

  - Question 3 is a reaction scheme where the carbon chain is increased leading to a carboxylic acid, then a pH and buffer question about that acid.

  - Question 4 is about the chemistry of iron, linking bonding, transition mental properties, Brønsted–Lowry pairs, tests for Fe^{3+} and redox titrations.

- To do well in this paper you must revise the entire specification (syllabus) and especially Topic 1.2. Do not become put off by this load as many of the Topics in A2 are extensions of those in AS. For example:

  - Topic 4.1 with 2.1 – Energetics
  - Topic 4.2 with 1.4, 1.6 and 1.7 – The Periodic Table
  - Topic 4.3 and 4.4 with 2.4 and 2.5 – Equilibrium
  - Topic 4.5, 5.3 and 5.5 with 2.2 – Organic chemistry
  - Topic 5.1 with 1.5 – Redox
  - Topic 5.4 with 2.3 – Kinetics

This leaves Topics 1.1 (Atomic structure), 1.2 (Formulae, equations and moles) and 1.3 (Structure and bonding) all of which are fundamental to chemistry and much of these topics will have been revised by your teacher during the A2 year.

Practice Test: Unit 6B (Synoptic)

Time allowed 1 hour 30 min

All questions are taken from previous Edexcel Advanced GCE questions.
The answers are on pages 141–142.

Section A

1   A fertiliser is known to contain ammonium sulphate, $(NH_4)_2SO_4$, as the only ammonium salt. This question concerns methods for the determination of the ammonium and sulphate ions.

  a A sample weighing 3.80 g was dissolved in water and the volume made up to 250 cm^3. To 25.0 cm^3 portions of this solution about 5 cm^3 (an excess) of aqueous methanal was added. The following reaction took place:

$$4NH_4^+(aq) + 6HCHO(aq) \rightarrow C_6H_{12}N_4(aq) + 4H^+(aq) + 6H_2O(l)$$

  The liberated acid was titrated directly with 0.100 mol dm^{-3} aqueous sodium hydroxide. The average volume required was 28.0 cm^3. Calculate the percentage of ammonium sulphate in the fertiliser.   **[5]**

  b In a second determination of the ammonium ion content, the same mass of fertiliser (i.e. 3.80 g) was treated with excess sodium hydroxide and heated. The ammonia liberated was passed into a known excess of hydrochloric acid. The unreacted hydrochloric acid was then titrated with standard aqueous sodium hydroxide.
  The calculation of the percentage composition of the fertiliser gave a value that was 5% lower than the value obtained by the method in **a**. Suggest reasons for this error other than those arising from the measurement of volumes.   **[2]**

  c To determine the sulphate ion concentration of the fertiliser, aqueous barium chloride was added in excess to the fertiliser solution. The precipitate produced was filtered off and dried by strong heating.
    i   Give the ionic equation, with state symbols, for the precipitation reaction.   **[2]**
    ii Suggest why the aqueous barium chloride was added in excess.   **[1]**

  d Many carbonates are also insoluble, and can be precipitated, dried and weighed in experiments similar to **c**. However the strong heating needed to drive off all the water can cause a problem in determining the mass of the carbonate precipitated. Suggest what this problem is, and, choosing a suitable example, write an equation for a reaction that might occur when the precipitate is heated.   **[4]**

(Total 14 marks)
[June 2002 Unit Test 6B question 1]

Section B

Answer TWO questions only

2 a The covalent compound urea, $(NH_2)_2C = O$, is commonly used as a fertiliser in most of the European Union, whereas in the UK the most popular fertiliser is ionic ammonium nitrate, NH_4NO_3. Apart from its nitrogen content suggest **two** advantages of using urea as a fertiliser compared with using ammonium nitrate.   **[2]**

  b The ammonium ion in water has an acid dissociation constant $K_a = 5.62 \times 10^{-10}$ mol dm^{-3}. The conjugate acid of urea has $K_a = 0.66$ mol dm^{-3}. Use this data to explain which of ammonia or urea is the stronger base.   **[2]**

  c Some organic nitrogen compounds are used to manufacture polyamides by condensation polymerisation. With the aid of diagrams, define the terms *condensation polymerisation and polyamide*.   **[4]**

  d Ethanamide, CH_3CONH_2, can be converted to methylamine, CH_3NH_2.
    i   State the reagents and conditions for carrying out this conversion.   **[3]**
    ii Suggest the formula of the likely product if urea were used instead of ethanamide in this conversion.   **[1]**

  e Ammonium nitrate can explode when heated strongly.

$$NH_4NO_3(s) \rightarrow N_2O(g) + 2H_2O(g) \qquad \Delta H = -23 \text{ kJ mol}^{-1}$$

With moderate heating the ammonium nitrate volatilises reversibly

$$NH_4NO_3(s) \rightleftharpoons NH_3(g) + HNO_3(g) \qquad \Delta H = +171 \text{ kJ mol}^{-1}$$

i State why the expression for K_p for the reversible change does not include ammonium nitrate. **[1]**

ii Explain the concepts of *thermodynamic* and *kinetic stability* with reference to these two reactions.

[5]

(Total 18 marks)

[June 2001CH6 question 4]

3 The principal source of the odour in sweat is the organic acid **HA**, $C_7H_{12}O_2$. It will decolourise bromine water immediately, and it shows geometric isomerism.

a **HA** can be made by the following sequence of reactions:

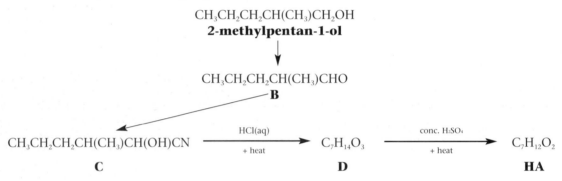

$CH_3CH_2CH_2CH(CH_3)CH_2OH$
2-methylpentan-1-ol

$CH_3CH_2CH_2CH(CH_3)CHO$
B

$CH_3CH_2CH_2CH(CH_3)CH(OH)CN \xrightarrow[\text{+ heat}]{\text{HCl(aq)}} C_7H_{14}O_3 \xrightarrow[\text{+ heat}]{\text{conc. H}_2\text{SO}_4} C_7H_{12}O_2$
C   **D**   **HA**

i Give the reagents and conditions for the oxidation of 2-methylpentan-1-ol to **B**. **[3]**

ii State the conditions and write the mechanism for the reaction of HCN with **B** to give **C**. **[4]**

iii Give the equation for the hydrolysis of **C** to **D**, and hence draw the structural formula of **D**. **[2]**

iv Draw the structure of **HA**. **[2]**

b **HA** has a dissociation constant $K_a = 4.50 \times 10^{-5}$ mol dm^{-3}, and is a monobasic (monoprotic) acid which dissociates according to the equation:

$$\mathbf{HA}(aq) + H_2O(l) \rightleftharpoons H_3O^+(aq) + \mathbf{A}^-(aq)$$

i Write the expression for K_a and calculate the pH of a solution of **HA** of concentration 0.0100 mol dm^{-3}. **[3]**

ii The pH of skin is almost constant. Explain how, in principle, this could be achieved by using **HA** and one of its salts. (Given its smell, it is perhaps fortunate that skin does not use this compound as a pH regulator) **[4]**

(Total 18 marks)

[January 2002 CH6 question 3]

4 Iron(III) chloride exists in two forms, the covalent anhydrous chloride and the ionic hydrated chloride.

a Explain why anhydrous iron(III) chloride is covalent whereas sodium chloride is ionic. **[4]**

b Hydrated iron(III) chloride is soluble in water and its solution is acidic.

i Write an ionic equation to show why its solution is acidic and identify the acid-base conjugate pairs in this reaction. **[3]**

ii Describe what you would see when a solution of sodium hydroxide is added until in excess to a solution of iron(III) chloride.

Write an ionic equation for this reaction and identify the type of reaction. **[4]**

iii A test for Fe^{3+} ions is to add a solution of thiocyanate ions, $SCN^-(aq)$, which react to form a blood red ion of formula $[FeSCN(H_2O)_5]^{2+}$. What is the bonding between the SCN^- ion and the Fe^{3+} ion, and what is the cause of the colour of the ion? **[3]**

c Iron also forms the intensely red coloured anion, FeO_4^{2-}. In acidic solution this is a powerful oxidising agent, and is itself reduced to Fe^{3+} ions.

Calculate the oxidation number of iron in FeO_4^{2-}. Suggest, in outline, how you could determine the concentration of a solution of FeO_4^{2-} ions. **[4]**

(Total 18 marks)

[January 2001 CH6 question 6]

Answers

Unit 1

Topic 1.1 Atomic structure

1 The relative **isotopic** mass is the mass of a single isotope of an element (relative to 1/12th the mass of a carbon 12 atom), whereas the relative **atomic** mass is the average mass taking into account the isotopes of the element. The relative isotopic masses of the two chlorine isotopes are 35 and 37, but the relative atomic mass of chlorine is 35.5

2 There are many more ^{7}Li atoms than ^{6}Li atoms in natural lithium, so the average is closer to 7.

3 A_r of magnesium

$$= \frac{(24.0 \times 78.6 + 25.0 \times 10.1 + 26.0 \times 11.3)}{100} = 24.3$$

4 Let the % of ^{69}Ga be x,
and so the % of ^{71}Ga = $(100 - x)$

$$69.8 = \frac{[69.0x + 71.0(100 - x)]}{100}$$

Therefore $6980 = 69x + 7100 - 71x$

Therefore $2x = 120$ and $x = 60.0$

Gallium contains 60.0% of the ^{69}Ga isotope.

5 Group 5. There is a big jump between the 5th and 6th ionisation energies. Therefore the 6th electron is being removed from a lower shell (nearer to the nucleus) than the first five electrons.

6 a

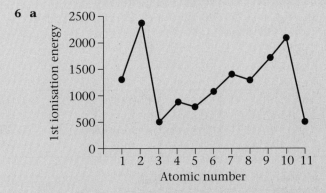

b i A 2p electron, which is in a higher energy level, is being removed. So less energy is required to remove it compared to that required to remove a 2s electron.

ii There are two electrons in the $2p_x$ orbital which repel one another, so less energy is required to remove one of the pair than if there had only been one electron in that orbital.

iii Although the element with Z = 11 has 8 more protons, it also has 8 more shielding electrons, but as the 3s electron is further from the nucleus than a 2s electron, so less energy is required to remove it.

7 The electron in a box notation for chlorine is:

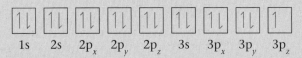

8 All have one electron in their outer shells, but K is bigger (larger atomic radius) than Na, which is bigger than Li. Therefore the outer electron is held on less firmly (the extra nuclear charge is balanced by the same extra number of shielding electrons) and so is the easiest to remove from potassium and the hardest to remove from Li. This makes potassium the most reactive.

9 a i $Li(g) + e^- \rightarrow Li^-(g)$

ii $Cl(g) + e^- \rightarrow Cl^-(g)$

iii $O(g) + e^- \rightarrow O^-(g)$

> ### Examiner's comments
>
> Note that the species on the left hand side is a gaseous **atom** and the one on the right is a **negative** ion.

b $O^-(g) + e^- \rightarrow O^{2-}(g)$

c For the 1st electron affinity, the (negative) electron is brought closer to the positive nucleus in a **neutral** atom, and so energy is released. For the 2nd electron affinity, the negative electron is brought towards a **negative** ion, and so repulsion has to be overcome, which requires energy.

Topic 1.2 Formulae, equations and moles

1 a

	%	÷ 6.90
C	$82.76 \div 12 = 6.90$	1
H	$17.24 \div 1 = 17.24$	2.5

Therefore the empirical formula is C_2H_5

b Mass of C_2H_5 is 29, but $M_r = 58$,
Therefore the molecular formula is C_4H_{10}

2 a $4NH_3 + 5O_2 \rightarrow 4NO + 6H_2O$

b $2Fe^{3+}(aq) + Sn^{2+}(aq) \rightarrow 2Fe^{2+}(aq) + Sn^{4+}(aq)$

3 Equation: $H_3PO_4 + 3NaOH \rightarrow Na_3PO_4 + 3H_2O$
   Amount of H_3PO_4 = 2.34/98 = 0.02388 mol
   Amount of NaOH = 0.02388 × 3/1 = 0.07163 mol
   Mass of NaOH = 0.07163 × 40 = 2.87 g

4 Equation: $2KOH + H_2SO_4 = K_2SO_4 + 2H_2O$
   Amount of H_2SO_4 = 0.0125 dm^3 × 0.0747 mol dm^{-3}
       = 9.338 × 10^{-4} mol
   Amount of KOH = 9.338 × 10^{-4} × 2/1
       = 1.868 × 10^{-3} mol
   Volume of KOH solution = mol/conc
       = $\dfrac{1.868 \times 10^{-3} \text{ mol}}{0.107 \text{ mol dm}^{-3}}$
       = 0.0175 dm^3 = 17.5 cm^3

5 Amount of $Bi(NO_3)_3$ = 0.025 dm^3 × 0.55 mol dm^{-3}
       = 0.0138 mol
   Amount of H_2S = 0.0138 × 3/2
       = 0.0206 mol
   Volume of H_2S gas = mol × molar volume
       = 0.0206 × 24 = 0.50 dm^3

6 Ratio mol of $H_2(g)$:mol of $CH_4(g)$ = 3:1
   Volume of $H_2(g)$:volume of $CH_4(g)$ = 3:1
   Volume of $H_2(g)$ = 3 × 33 = 99 dm^3

Topic 1.3 Structure and bonding

1

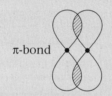

σ-bond   π-bond

2 a $AlCl_3$ because **either** there is less difference in electronegativity between Al and Cl than between Al and F
   or Al^{3+} is very polarising and the bigger Cl^- ion is more polarisable than the smaller F^- ion.

  b $BeCl_2$ because **either** there is less difference in electronegativity between Be and Cl than between Mg and Cl
   or Be^{2+} is smaller than Mg^{2+} and so is more polarising.

3

	a Hydrogen bond	**b** Dispersion	**c** Dipole /dipole
HF	Yes	Yes	Yes
I_2	No	Yes	No
HBr	No	Yes	Yes
PH_3	No	Yes	Yes
Ar	No	Yes	No

4 iodine   **e** simple molecular
  ice   **d** H bonded molecular

copper   **a** metallic
SiO_2 **b** giant atomic
solid CO_2 **e** simple molecular
CaO   **c** ionic
poly(ethene)   **f** polymeric
graphite   **b** giant atomic
sulphur   **e** simple molecular
$CuSO_4$ **c** ionic
sucrose   **d** H-bonded molecular
LiF   **c** ionic

5 Melting: on heating, the molecules or ions vibrate **more** until the intermolecular forces (in molecular) or ion/ion forces (in ionic) substances are overcome and the lattice breaks down.

Boiling: on heating the average kinetic energy of the molecules increases. Those with a large enough kinetic energy and hence speed escape, not only from the surface but also from the body of the liquid, forming bubbles.

6 a NH_3. There are stronger **inter**molecular forces in NH_3 as it forms H bonds and PH_3 does not.

  b HBr. It has more electrons in its molecule, and so it has stronger dispersion forces (which outweigh the difference in dipole/dipole forces).

  c Propanone. Both have about the same strength of dispersion forces because they have about the same number of electrons, but propanone is polar and so has dipole/dipole forces as well.

  d S_8. It has about twice as many electrons as P_4, and so has stronger dispersion forces.

  e NaCl. It has ion/ion forces that are very much stronger than the weak intermolecular dispersion forces between CCl_4 molecules.

7

	Number of σ bond pairs	Number of lone pairs	Total number of electron pairs	Shape
SiH_4	4	0	4	Tetrahedral
BF_3	3	0	3	Triangular planar
$BeCl_2$	2	0	2	Linear
PCl_3	3	1	4	Pyramidal
SF_6	6	0	6	Octahedral
XeF_4	4	2	6	Square planar
NH_4^+	4	0	4	Tetrahedral
PCl_6^-	6	0	6	Octahedral

8 In Na (s) there are loosely held delocalised electrons that can move through the lattice. In

NaCl(s) the electrons are localised on the ions, not loosely held **and** the ions are fixed in their positions in the lattice and so are not free to move.

Topic 1.4 The Periodic Table I

1 a They all have the same number of electrons in their outer shell. For Group 2 this number is 2.

b This period is Period 4, and so all the members have electrons in four shells.

2 a H_2 is simple molecular
b Na is metallic
c Si is giant atomic
d S_8 is simple molecular
e Cl_2 is simple molecular
f Ar is simple molecular (argon is monatomic).

Examiner's comments

The answer 'hydrogen bonded' or 'ionic' cannot apply to elements.

3 Sodium and magnesium are metals. Magnesium is 2+ and has a smaller ionic radius, therefore it will have stronger metallic bonds and a higher melting temperature. Silicon forms a giant atomic structure with each silicon atom joined covalently to four other silicon atoms. These strong covalent bonds have to be broken. This requires a huge amount of energy and so the melting temperature is very high.

Examiner's comments

Carbon, in the form of diamond, has a similar structure, but as the smaller carbon atom forms stronger bonds than silicon, diamond's melting temperature is even higher.

White phosphorus, sulphur and chlorine all form simple molecular solids. The melting temperatures depend upon the strengths of their intermolecular forces which are dispersion (induced dipole/induced dipole) forces. The strength of this force depends upon the number of electrons in the molecule.

Examiner's comments

The simplest way in which to work this out is to add together the atomic numbers of all the atoms in the molecule.

P_4 has $4 \times 15 = 60$, S_8 has $8 \times 16 = 128$ and Cl_2 has $2 \times 17 = 34$ electrons per molecule. Thus sulphur has the strongest intermolecular forces and the highest melting temperature and chlorine has the weakest intermolecular forces and the lowest melting temperature.

4 a i Nuclear charge is equal to the number of protons in the nucleus.

ii Screening or shielding occurs when an outer electron is insulated from the pull of the positive nucleus by electrons in inner shells.

b i The 1st ionisation energy of sodium is for the removal of the 3s electron. Sodium has a nuclear charge of +11, but the 3s electron is shielded by the 10 electrons in the 1st and 2nd shells. The second electron that is removed comes from the 2nd shell (it is a 2p electron). This electron is shielded from the nucleus only by electrons which are in a lower shell, which in this case is by the two 1s electrons and so it is held much more strongly by the nucleus. It is also nearer to the nucleus, which makes it even harder to remove.

ii Sodium has a nuclear charge of +11 and its 3s electron is shielded by the inner 1st and 2nd shell electrons. Magnesium has a nuclear charge of +12 (one more than sodium), but its 3s electron is shielded by the same inner electrons. Thus its 3s electrons are pulled more strongly to the nucleus and are harder to remove.

Examiner's comments

It is because of this extra pull that a magnesium atom is smaller than a sodium atom.

Topic 1.5 Introduction to oxidation and reduction

1 a $Sn^{2+}(aq) \rightarrow Sn^{4+}(aq) + 2e^-$

Examiner's comments

Electrons are on right because Sn^{2+} is oxidised: 2e- is needed because oxidation number changes by 2.

b $Fe^{3+}(aq) + e^- \rightarrow Fe^{2+}(aq)$
c $Sn^{2+}(aq) + 2Fe^{3+}(aq) \rightarrow Sn^{4+}(aq) + 2Fe^{2+}(aq)$

Examiner's comments

Equation **b** has to be doubled, so that the electrons cancel when 2 times **b** is added to **a**.

2 a $H_2O_2(aq) + 2H^+(aq) + 2e^- \rightarrow 2H_2O(l)$

Examiner's comments

H_2O_2 is reduced, so electrons are on the left.

b $S(s) + 2H^+(aq) + 2e^- \rightarrow H_2S(aq)$

> ### Examiner's comments
>
> $2e^-$ are needed on the left because S is reduced and its oxidation number goes from 0 to –2.

c $H_2O_2(aq) + H_2S(aq) \rightarrow 2H_2O(l) + S(s)$

> ### Examiner's comments
>
> Both H_2S and H_2O_2 must be on the left because they are the reactants; equation **b** is reversed and added to **a**; note that the H^+ ions cancel out.

3 a $PbO_2(s) + 2H_2SO_4(aq) + 2e^- \rightarrow PbSO_4(s) + 2H_2O(l) + SO_4^{2-}(aq)$

> ### Examiner's comments
>
> $2e^-$ are needed on the left as PbO_2 is reduced and its oxidation number changes from +4 to +2.

b $PbSO_4(s) + 2e^- \rightarrow Pb(s) + SO_4^{2-}(aq)$

> ### Examiner's comments
>
> $2e^-$ are needed on the left because $PbSO_4$ is reduced and the oxidation number of lead changes from +2 to 0.

c $PbO_2(s) + Pb(s) + 2H_2SO_4(aq) \rightarrow 2PbSO_4(s) + 2H_2O(l)$

> ### Examiner's comments
>
> Because Pb is a reactant, equation **ii** has to be reversed and then added to **i**. This is the reaction that takes place when current is drawn from a lead-acid battery.

Topic 1.6 Group 1 and Group 2

1 Dip a hot platinum (or nichrome) wire in clean concentrated HCl and then into the solid. Place in a hot bunsen flame. The substance that turns the flame carmine red is LiCl, the one which turns the flame lilac is KCl and the one that turns the flame apple green is $BaCl_2$.

2 The heat of the flame vaporises the sodium chloride, producing some Na and Cl **atoms**. Electrons are promoted into the 4th shell in some of the sodium atoms. These electrons then fall back to the ground state, which is the 3s orbital, and energy is given out in the form of light. The flame

is yellow because the energy difference between the 4th shell and the 3s orbital is equivalent to the yellow line in the spectrum.

3 The sodium atom is smaller than the potassium atom, because sodium has three shells of electrons and potassium has four. Thus sodium's outer s electron is held on more firmly, and more energy is required to remove it.

> ### Examiner's comments
>
> This is the case in spite of the fact that potassium has eight more protons than sodium, because it also has eight more electrons between the nucleus and the outer electron, so the extra nuclear charge is balanced by the extra shielding.

4 a $2Ca + O_2 \rightarrow 2CaO$ (or $Ca + {}^1\!/_2O_2 \rightarrow CaO$)
  b $Ca + 2H_2O \rightarrow Ca(OH)_2 + H_2$
  c $2K + 2H_2O \rightarrow 2KOH + H_2$
  d $Mg(s) + H_2O(g) \rightarrow MgO(s) + H_2(g)$

5 The white precipitate is the insoluble magnesium hydroxide. Barium hydroxide is soluble and so is not precipitated.

6 Beryllium, as $BeCO_3$ is the least thermally stable of the Group 2 carbonates. It decomposes very easily on heating because the Be^{2+} ion is very polarising. The other Group 2 ions are less polarising because they are larger.

7 a $4LiNO_3 \rightarrow 2Li_2O + 4NO_2 + O_2$
  $2NaNO_3 \rightarrow 2NaNO_2 + O_2$
  $2Mg(NO_3)_2 \rightarrow 2MgO + 4NO_2 + O_2$
  b $Na_2CO_3 \rightarrow$ no reaction
  $MgCO_3 \rightarrow MgO + CO_2$
  $BaCO_3 \rightarrow$ no reaction

Topic 1.7 Group 7

1 Covalent HCl ionises when added to water, and the hydration energy given out by the ions bonding with the water molecules is greater than the energy required to break the covalent H–Cl bonds:

$$HCl(g) + H_2O(l) \rightarrow H_3O^+(aq) + Cl^-(aq)$$

2 First a solution has to be made. The sample can be dissolved either in water or in dilute nitric acid. (If water is chosen, dilute nitric acid must then be added to destroy the carbonate). To this slightly acidic solution, add silver nitrate solution. If there is chloride present, there will be a white precipitate that will dissolve in dilute aqueous ammonia.

3 H_2SO_4 is a **stronger acid** than HCl (or HI), and so it will protonate the Cl^- ions in the sodium chloride producing HCl gas. HCl is **not** a strong enough **reducing agent** to reduce sulphuric acid, and so no further reaction will take place. With sodium iodide, HI is produced by the protonation of the iodide ions, but the HI is a very strong reducing agent, and so it will reduce the sulphuric acid and itself become oxidised to iodine.

4 Disproportionation is a reaction in which an element is simultaneously oxidised and reduced. Chlorine disproportionates when added to alkali:

$$Cl_2 + 2OH^- \rightarrow Cl^- + ClO^- + H_2O$$
$$(0) \qquad (-1) \quad (+1)$$

Also, chlorate(I) disproportionates when heated:

$$3ClO^- \rightarrow 2Cl^- + ClO_3^-$$
$$(+1) \qquad (-1) \quad (+5)$$

Answers to Practice Test Unit 1

The allocation of marks follows Edexcel mark schemes.
The marks that you will need for each grade are approximately:

A	70%
B	61%
C	52%
D	44%
E	36%

1 a i $2Ca + O_2 \rightarrow 2CaO$ = **[1]**
　　ii $Na_2O + H_2O \rightarrow 2NaOH$ = **[1]**
　　iii $Na_2O + 2HCl \rightarrow 2NaCl + H_2O$
　　　　Species **[1]**, balancing **[1]** = **[2]**
　b The thermal stability increases down the group **[1]** as the cation size increases **[1]** and so the carbonate ion is distorted (or polarised) less **[1]** = **[3]**

2 a i Graphite has a giant atomic (or giant covalent) structure **[1]**. The bonding between atoms is covalent **[1]**.
Your diagram must show several interlinked hexagons **[1]** in layers **[1]** = **[4]**

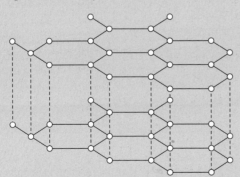

　　ii Sodium chloride has a lattice structure **[1]**. The bonding is ionic **[1]**.
Your diagram must show a 3-D arrangement of ions, Na^+ alternating with Cl^-. **[1]** = **[3]**

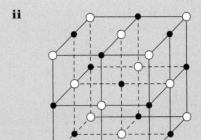

　b Graphite has covalent bonds between atoms that have to be broken **[1]**; sodium chloride has electrostatic forces between the ions **[1]**. In both these forces are strong and so a lot of energy is needed to break them **[1]** = **[3]**
　c i Graphite has delocalised electrons above and below the plane of hexagons **[1]**, which can flow (which are mobile) **[1]** = **[2]**
　　ii The lattice is broken down in the liquid and so the ions are free to move = **[1]**

3 a i Atomic number is the number of protons in the nucleus **[1]** Mass number is the total (or sum of the) number of protons and neutrons **[1]** = **[2]**
　　ii $^{79}_{35}Br^+$ = **[1]**
　b i High energy electrons bombard the Br_2 molecules **[1]** and knock out an electron **[1]** = **[2]**
　　ii They are accelerated by charged plates (or across an electric field) = **[1]**
　　iii $^{79}Br-^{81}Br^+$ = **[1]**

4 a
Divide each % by their A_r values **[1]** and then divide by the smallest **[1]**
　　H　　$11.1 \div 1 = 11.1$　　then $\div 7.4 = 1.5$
　　C　　$88.9 \div 12 = 7.4$　　then $\div 7.4 = 1$
　　Multiply both by 2 to get whole numbers
　　∴ E.F. is C_2H_3 **[1]** = **[3]**

b HI has more electrons **[1]** ∴ induced dipole
or dispersion or London or v.d.Waals
are stronger **[1]**   = **[2]**

c i 3 bond pairs and 1 lone pair repel to get
as far apart as possible **[1]** see diagram
below **[1]**   = **[2]**

ii 4 bond pairs around Al repel to get as
far apart as possible **[1]** see diagram
below **[1]**   = **[2]**

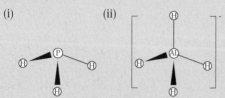

(i)   (ii)

d Moles of gas = volume ÷ molar volume
= 8.0 cm³/2.4 × 10⁴ cm³ mol⁻¹ = 3.33 × 10⁻⁴ mol
[1]
Number of molecules = 6.0 × 10²³ × number of
moles = 2.0 × 10²⁰ **[1]**   = **[2]**

$$= 8.0 \text{ cm}^3/2.4 \times 10^4 \text{ cm}^3 \text{ mol}^{-1} = 3.33 \times 10^{-4} \text{ mol}$$
$$= 6.0 \times 10^{23} \times \text{ number of}$$
$$= 2.0 \times 10^{20}$$

Examiner's comments

b Do not say because the molar mass of HI is larger.
c Remember that it is the electron pairs not the
atoms nor the bonds that repel to give maximum
separation.

---

5 a i The oxidation number of sulphur: in H_2SO_4 is
+6 **[1]**, in H_2S is +2 **[1]**, in SO_2 is +4 **[1]** = **[3]**

ii The I⁻ ion is the stronger reducing agent
because it reduces sulphur from +6 to +2 **[1]**,
whereas Br⁻ only reduces it to +4 **[1]**   = **[2]**

b i $2Cl^- \rightarrow Cl_2 + 2e^-$ or $Cl^- \rightarrow \frac{1}{2}Cl_2 + e^-$ = **[1]**

ii $OCl^- + 2H^+ + 2e^- \rightarrow Cl^- + H_2O$
species **[1]**, balance **[1]**   = **[2]**

iii $OCl^- + 2H^+ + Cl^- \rightarrow Cl_2 + H_2O$ = **[1]**

Examiner's comments

b i & ii For an oxidation reaction, the electrons are
on the right, and for reduction they are on the left.
iii You would not be penalised for failing to cancel
one of the Cl⁻ ions, i.e. having the equation.
$OCl^- + 2H^+ + 2Cl^- \rightarrow Cl_2 + H_2O + Cl^-$

---

6 a 2, 8, 6   = **[1]**

b i $S(g) + e^- \rightarrow S^-(g)$ **[1]**

ii $S(g) \rightarrow S^+(g) + e^-$ **[1]**
and **[1]** for state symbols in both equations
= **[3]**

c A big jump in ionisation energy means that an
electron has been removed from a new shell **[1]**
The first big jump is after 6 electrons and so
there are six electrons in the outer shell **[1]**
The second big jump is after 8 electrons and so
there are 8 in the 2nd shell **[1]**   = **[3]**

---

d i **[1]** for the electrons around S and **[1]** for the
electrons around both Cl atoms.   = **[2]**

ii The 2 bond pairs and the 2 lone pairs **[1]**
repel each other to get as far apart as
possible **[1]**   = **[2]**

iii Cl is more electronegative than S, so the
bonds are polar **[1]**, but as the molecule is
bent, the polarities do not cancel **[1]**   = **[2]**

Unit 2

Topic 2.1 Energetics I

1 Gases at a pressure of 1 atmosphere, a stated
temperature (298 K) and all solutions of
concentration 1.00 mol dm⁻³. Substances in their
most stable state.

2 i $2C(s) + 2H_2(g) + O_2(g) \rightarrow CH_3COOH(l)$

ii $CH_3COOH(l) + 2O_2(g) \rightarrow 2CO_2(g) + 2H_2O(l)$

iii $CH_3COOH(aq) + NaOH(aq) \rightarrow CH_3COONa(aq)$
$+ H_2O(l)$
or $CH_3COOH(aq) + OH^-(aq) \rightarrow CH_3COO^-(aq)$
$+ H_2O(l)$

Examiner's comments

Note that ethanoic acid is a weak acid, and so its
formula must be written in full and not as H⁺.

3

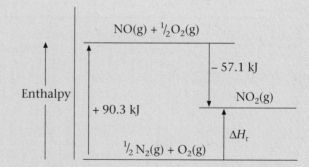

$\Delta H_r = +90.3$ kJ + (−57.1) kJ = +33.2 kJ mol⁻¹

Examiner's comments

Note that the + sign is given in the answer in order
to emphasise that the reaction is endothermic.

4

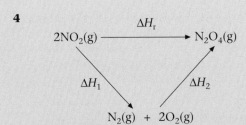

$\Delta H_r = \Delta H_1 + \Delta H_2 = {}^-2 \times (+33.9) + (+9.7) = {}^-58.1$ kJ mol^{-1}

Examiner's comments

ΔH_1 = **minus** 2 **times** the enthalpy of formation of $NO_2(g)$, because the reaction represents the formation of $NO_2(g)$ **reversed** and ΔH_2 = enthalpy of formation of $N_2O_4(g)$.

5

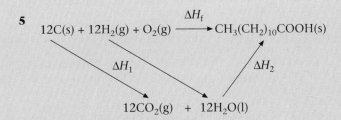

where ΔH_1 = 12 × enthalpy of combustion of carbon (graphite) + 12 × enthalpy of combustion of hydrogen and
ΔH_2 = **minus** the enthalpy of combustion of lauric acid.
ΔH_f = 12 × (−394) + 12 × (−286) − (−7377) = −783 kJ mol^{-1}.

6 HCl + NaOH → NaCl + H$_2$O

Amount of HCl = conc(mol dm^{-3}) × vol(dm^3)

$$= 1.00 \times \frac{100}{1000}$$

$$= 0.100 \text{ mol}$$

Rise in temperature = +6.80 °C
Heat produced when 0.100 mol reacts
= total mass × specific heat capacity × rise in temperature
= 200 g × 4.18 J g^{-1} °C^{-1} × 6.80 °C = 5685 J
= 5.685 kJ
Heat produced when 1 mol reacts
= 5.685/0.100 = 56.9 kJ mol^{-1}
therefore ΔH_{neut} = −56.9 kJ mol^{-1} (exothermic).

Examiner's comments

Because the mixture became hotter, the reaction is exothermic. The mass used is the mass of the whole solution, 100 cm^3 + 100 cm^3 = 200 cm^3 = 200 g, not the mass of the HCl.

7

Break (endo)
1 × C=C = +612 kJ
1 × H–O = +463 kJ

Make (exo)
1 × C–C = −348 kJ
1 × C–H = −412 kJ
1 × C–O = −360 kJ

Total   = +1075 kJ

Total   = −1120 kJ

ΔH = net result of bonds broken and bonds made
= +1075 − 1120 = −45 kJ mol^{-1}.

Topic 2.2   Organic chemistry I

1 a

Br—C—C—C—H with Br, F, F on top and Cl, Cl, H on bottom

b

H—C—C—C—C—H with Cl, OH, H, H on top and H, H, H, H on bottom

2 a CH$_3$CH$_2$CH$_2$OH and CH$_3$CH(OH)CH$_3$.

Examiner's comments

This type of unambiguous formula is allowed unless a **full** structural formula is asked for.

b CH$_3$CH$_2$CH$_2$CH$_2$CH$_3$, CH$_3$C(CH$_3$)$_2$CH$_3$ and CH$_3$CH$_2$CH(CH$_3$)CH$_3$
c CH$_3$CH=CHCH$_3$ (both the cis and trans isomers), (CH$_3$)$_2$C=CH$_2$ and CH$_3$CH$_2$CH=CH$_2$

3 a A free radical is a species with a single unpaired electron, e.g. Cl·.
b Homolytic fission is when a bond breaks with one electron going to each atom.

4 a An electrophile is a species which seeks out negative centres and accepts a lone pair of electrons to form a new covalent bond.
b Heterolytic fission is when a bond breaks with the two electrons going to one atom.

5 a C$_2$H$_6$ + Cl$_2$ → C$_2$H$_5$Cl + HCl
then: C$_2$H$_5$Cl + Cl$_2$ → C$_2$H$_4$Cl$_2$ + HCl etc
conditions: ultraviolet light
b C$_2$H$_6$ + 3^1/$_2$O$_2$ → 2CO$_2$ + 3H$_2$O
conditions: flame or a spark

6 a CH$_3$CH=CH$_2$ + H$_2$ → CH$_3$CH$_2$CH$_3$
conditions: heat and Pt catalyst
b CH$_3$CH=CH$_2$ + Br$_2$ → CH$_3$CHBrCH$_2$Br
conditions: solution in hexane
observation: goes from brown to colourless
c CH$_3$CH=CH$_2$ + HI → CH$_3$CHICH$_3$
conditions: mix gases at room temperature
d CH$_3$CH=CH$_2$ + [O] + H$_2$O → CH$_3$CH(OH)CH$_2$OH
condition: aqueous
observation: goes from purple to brown precipitate

7 a $CH_3CHICH_3 + NaOH \rightarrow CH_3CH(OH)CH_3 + NaI$
propan-2-ol
  b $CH_3CHICH_3 + KOH \rightarrow CH_3CH=CH_2 + KI + H_2O$
propene
  c $CH_3CHICH_3 + 2NH_3 \rightarrow CH_3CH(NH_2)CH_3 + NH_4I$
2-aminopropane

8 $CH_3CH(OH)CH_3 - H_2O \rightarrow CH_3CH=CH_2$
conditions: concentrated acid at 170 °C

9 a

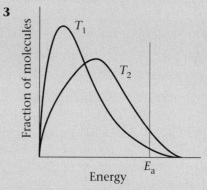

Propanoic acid
observation: solution goes from orange to green
  b No reaction
observation: solution stays orange
  c $CH_3CHClCH_3$
2-chloropropane
observation: steamy fumes

10 The π carbon–carbon bond in ethene is weaker than the σ carbon–carbon bond in ethane. This makes the activation energy less, and so the reaction is faster.

11 The C–Br bond is stronger than the C–I bond. This makes the activation energy more, and so the reaction is slower.

12 $C_6H_{10} + Br_2 \rightarrow C_6H_{10}Br_2$
5.67 g of C_6H_{10} = 5.67/82 = 0.0691 mol
Because the reaction is a 1:1 reaction, the amount of product is also 0.0691 mol.
Therefore the theoretical yield of $C_6H_{10}Br_2$
= 0.0691 × 242 = 16.7 g
The % yield = $\frac{15.4 \text{ g}}{16.7 \text{ g}} \times 100$ = 92%.

13 They are stable in air owing to the strong C–Cl and C–F bonds.
They then diffuse into the stratosphere where they form radicals and destroy the ozone layer.

Topic 2.3 Kinetics I

1 i The activation energy is the minimum energy that the reacting molecules must have when they collide in order for them to react. It is measured in $kJ \, mol^{-1}$.
  ii A catalyst is a substance that speeds up a chemical reaction without being used up. It works by providing an **alternative** route with a smaller activation energy.

2 Pressure. As the pressure is increased, the number of molecules in a given volume increases, and the frequency of collision increases and hence the rate of reaction, which is dependent upon this frequency of collision, increases too.

Temperature. If the temperature is increased, the average kinetic energy of the molecules is also increased. This means that a greater proportion of the colliding molecules will have, between them, energy greater than or equal to the activation energy. There will be a larger number of **successful** collisions and a faster rate of reaction.

Catalyst. This increases the rate of reaction by providing an alternative path with a lower activation energy, thus a greater proportion of the colliding molecules have energy greater than or equal to the lower catalysed activation energy.

3

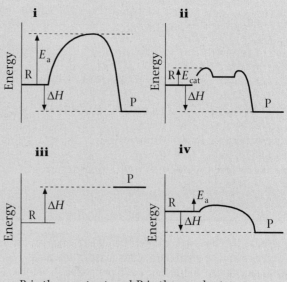

4 Bacterial and enzyme reactions are also slowed down by a decrease in temperature for exactly the same reason as ordinary chemical reactions, which is that fewer colliding molecules will have the necessary activation energy to react on collision. So the low temperature in the refrigerator will slow down the biological decay reactions.

5

R is the reactant and P is the product.

Topic 2.4 Chemical equilibria I

1 i true
   ii false
   iii true
2 i Position moves R→L, because that is the exothermic direction.
   ii Pressure increases, therefore position moves R→L as fewer gas moles on left
   iii None, catalyst speeds up the forward and reverse reactions equally, reducing the time taken to reach equilibrium.
3 i Alkali removes ethanoic acid, therefore position moves R→L.
   ii No change because ΔH is zero
4 Solid dissolves because high $[Cl^-]$ drives equilibrium L→R.

5 Too high a temperature produces too low a yield.
   Too low a temperature produces too slow a rate
   Catalyst allows reaction to proceed at a fast rate at a temperature where the yield is high.

Topic 2.5 Industrial inorganic chemistry

1 Reaction is slow at moderate temperatures. Therefore a catalyst is used which allows a fast rate and reasonable yield at 400 °C, that is, an economic rate and yield. A very high pressure (200 atm) is used, even though this is very expensive because the yield at ordinary pressures is very low. A very high pressure drives the equilibrium L→R to the side with fewer gas moles. After NH_3 is separated, the unreacted gases are recycled.

2 The ammonia liquefies under the high pressure and is removed from the unreacted nitrogen and hydrogen, which are then put back through the reaction chamber.

3 Anode: $2Cl^-(aq) \rightarrow Cl_2(g) + 2e^-$
   Cathode: $2H_2O(l) + 2e^- \rightarrow 2OH^-(aq) + H_2(g)$
   or $2H^+(aq) + 2e^- \rightarrow H_2(g)$

Answers to Practice Test Unit 2

The allocation of marks follows Edexcel mark schemes.
The marks that you will need each grade are approximately:

A	70%
B	61%
C	52%
D	44%
E	36%

1 a It is a series of compounds with the same general formula [1] which differ by CH_2 [1], and have the same functional group [1]   = [3]
   b $-(CH(CH_3)CH_2)_n-$: correct carbon chain [1] with extension bonds at both ends [1]   = [2]
   c i See Fig 2.4 on page 38. [1] for each isomer   = [2]
      ii Rotation around the π bond is restricted = [1]

Examiner's comments

b Do not have the 3 carbon atoms in a straight chain.
c Do not draw the structure with a bond angles of 90° around the C=C group.

2 a See Fig 2.9 on page 45: [1] for having the axes correct (y-axis labelled fraction of molecules, x-axis energy), [1] for having the curves with the correct shape, and [1] for having the T_2 curve flatter with the peak moved to the right   = [3]
   b Draw the activation energy on the graph to the right of both peaks [1], explanation that the area under the curve to the right of the E_a line measures the number of molecules with $E \geqslant E_a$ [1], which is larger at T_2 than at T_1[1], so there are a greater proportion of successful collisions [1]   = [4]

Examiner's comments

a The total area under the two curves should be the same.
b Mention must be made of collisions.

3 a i The reagent is HBr [1] and the conditions are in the gas phase [1]   = [2]
      ii **S** is $CH_3CHBrCH_3$ = [1]
   b i **P** is $CH_3CH(OH)CH_3$ = [1]
      ii The type of reaction in step 1 is electrophilic [1] addition [1]   = [2]
      in the conversion of **S** to **P** the type is nucleophilic [1] substitution [1]   = [2]
   c i For step 2, the reagent is **conc** sulphuric (or phosphoric) acid or aluminium oxide [1] and the condition is heat [1]   = [2]
      ii For step 3, the reagents are potassium dichromate(VI) [1] and sulphuric acid [1] and the conditions are heat under reflux [1] = [3]
   d **Q** is $CH_3CH(OH)CH_2OH$ [1]   = [1]

4 a Hess's law states that the enthalpy change for a reaction **[1]** is independent of the route **[1]** = **[2]**

b The definition is the enthalpy (or heat but **not** energy) change when 1 mole **[1]** of a substance is completely burnt in oxygen (or burnt in excess oxygen) **[1]** under standard conditions (at 1 atm pressure) **[1]** = **[3]**

c i
Bonds broken		bonds made	
$1 \times$ C–C	= + 348	$4 \times$ C=O	= –2972
$5 \times$ C–H	= + 2060	$6 \times$ O–H	= –2778
$1 \times$ C–O	= + 360	Total	= –5750 kJ **[1]**
$1 \times$ O–H	= + 463		
$3 \times$ O=O	= + 1488		
Total	= + 4719 kJ **[1]**		

ΔH = + 4719 – 5750 = –1031 kJ mol^{-1} **[1]** = **[3]**

ii See Fig. 2.1 on page 33.
Labelling of reactants and products **[1]**, with energy of reactants higher than products and ΔH shown **[1]** = **[2]**

5 a i Your answer for the temperature should be between 350 and 500 °C (623 and 773K) = **[1]**

ii A high temperature favours a high rate of reaction **[1]**, but gives a low yield **[1]**, so a temperature is chosen which gives a balance between rate and yield **[1]** = **[3]**

iii The catalyst used is iron = **[1]**

iv A catalyst provides an alternative pathway **[1]**, which has a lower activation energy **[1]** = **[2]**

b i Homogeneous means that all the substances are in the same phase (here all are in the gaseous state) = **[1]**

ii Dynamic means that the reaction is occurring in both directions **[1]** and equilibrium means that there is no change in concentrations **[1]** = **[2]**

c i A lower temperature will move the position to the right **[1]**, because that is the exothermic direction (or this produces heat) **[1]** = **[2]**

ii Decreasing the volume increases the pressure and so moves the equilibrium to the right **[1]** as the right has fewer **gas** molecules **[1]** = **[2]**

6 a $2O^{2-} \rightarrow O_2 + 4e^-$ (or $O^{2-} \rightarrow \frac{1}{2}O_2 + 2e^-$) = **[1]**

b $Al^{3+} + 3e^- \rightarrow Al$ = **[1]**

c The reaction at the anode. = **[1]**

d The oxygen produced at the anode reacts with the carbon anode and eats it away = **[1]**

e Aluminium oxide (pure bauxite) has an extremely high melting temperature **[1]**, and so the process would be too expensive **[1]** = **[2]**

Answers to Practice Test Unit 3B

The marks that you will need for each grade are approximately:

A	70%
B	62%
C	54%
D	47%
E	40%

1 Hydrogen: burns with a squeaky pop **[1]**
Oxygen: relights splint **[1]**
Carbon dioxide turns lime-water **[1]** milky **[1]**
Sulphur dioxide: the solution turns green **[1]**
The gas which turns moist litmus red then bleaches it is chlorine **[1]** = **[6]**

2 a i There is a loss in mass because CO_2 gas **[1]** is lost from the flask **[1]** = **[2]**

ii At first the slope of the graph is steep, so the reaction is fast, then as the slope gets less it

slows down **[1]**, when the line is horizontal (after 6 or 7 minutes) the reaction has stopped (all the acid has been used up) **[1]**
= **[2]**

b For experiment 2: your graph should initially be steeper than in 1 but is horizontal earlier, but with the same mass loss (1 g) as in 1. **[1]**
For experiment 3: your graph should initially be less steep. It should become horizontal later, but at the same mass loss as in 1. **[1]**
For experiment 4: your graph should be steeper than in 1 and become horizontal at twice the mass loss (2 g) as in 1. **[1]**   = **[3]**

c i Amount of HCl = 1.00 mol dm^{-3} × 50/1000 dm^3 = 0.0500 mol **[1]**, amount of CaCO3 = $\frac{1}{2}$ × 0.0500 mol = 0.0250 mol **[1]**:   mass of CaCO$_3$ = 0.0250 mol × 100 g mol^{-1} = 2.50 g **[1]**   = **[3]**

ii As the amount of acid in experiment 4 is twice that of the other experiments **[1]**, the mass of CaCO$_3$ must be at least 2 × 2.50 = 5.00 g **[1]**   = **[2]**

Examiner's comments

a To answer this you have to work out if the reaction is faster or slower than in 1 and whether the mass of CO$_2$ lost is the same, more or less than in 1.

c i You have to halve the moles of HCl because there are half as many moles of CaCO$_3$ as moles of HCl.

---

3 a Any two of:
use a pipette to measure out the CuSO$_4$ solution **[1]** as the volume is measured more accurately; **[1]** use smaller pieces of iron **[1]** as there will be less heat loss in a faster reaction **[1]**;
use a polystyrene (or plastic) cup **[1]** as this reduces heat loss **[1]**;
measure temperature for several minutes before addition of iron **[1]** as this allows for a more accurate determination of the initial temperature **[1]**;
measure temperature more frequently **[1]** as this makes the extrapolation more accurate **[1]**;
use a thermometer which reads to 0.5 or 0.1 °C **[1]** as this gives a more accurate temperature **[1]**;
stir the reaction mixture gently **[1]** as this ensures an even temperature in the liquid **[1]**.
= **[4]**

b i Heat given out = 50.0g × 4.18Jg^{-1}deg^{-1} × 15.2 °C = 3180 J or 3.18 kJ   = **[1]**

ii Amount of CuSO$_4$ = 0.500 mol dm^{-3} × 50/1000 dm^3 = 0.025 mol   = **[1]**

iii −3.18 kJ ÷ 0.025 mol = −127 kJ mol^{-1} = **[2]**
(1mark for the negative sign and 1 for the number)

Examiner's comments

b The temperature goes up, so heat is produced, and the reaction is exothermic; $\therefore$ ΔH is negative.

---

4 a i Any one of:
use a larger beaker **[1]** as this will prevent acid spray escaping **[1]**;
powder the limestone **[1]** as this speeds up the reaction **[1]**;
use more acid **[1]** as this gives larger and hence more accurate titres **[1]**.   = **[2]**

ii Choose titres 1 and 3 **[1]**;
mean = $\frac{1}{2}$ (14.90 + 14.85) = 14.875 cm^3 **[1]**
= **[2]**

b i Amount of NaOH in titre = 0.100 mol dm^{-3} × 14.875/1000 dm^3 = 0.0014875 mol   = **[1]**

ii Amount of HCl in portion = 0.001488mol
= **[1]**

iii Total amount of HCl remaining = 10 × 0.001488 = 0.01488 mol   = **[1]**

iv Original amount of HCl = 2.0 mol dm^{-3} × 50/1000 dm^3 = 0.100 mol   = **[1]**

v Amount of HCl used = 0.100 − 0.01488 = 0.08512 mol   = **[1]**

vi Amount of CaCO$_3$ = $\frac{1}{2}$ × 0.08512 = 0.04256 mol **[1]** = 0.04256 mol × 100 g mol^{-1} = 4.256g **[1]**   = **[2]**

vii Purity = 100 × 4.256 g/5.24 g = 81.2%   = **[1]**

c i C which is between 14.75 and 14.95   = **[1]**

ii One of:
not reading from bottom of meniscus, not reading meniscus at eye level, burette not vertical, air bubble below tap, funnel left in top of burette   = **[1]**

Examiner's comments

a i If possible titration experiments should be designed to give titres of about 25 cm^3.
ii You should only choose titration values that are less than 0.2 cm^3 apart.
b vi Don't forget that the ratio of HCl to CaCO$_3$ is 2 : 1.

---

5 a Place the chemicals in a round bottomed flask **[1]**, fitted with a reflux condenser **[1]**. Heat for several minutes **[1]**. Distil off the propan-2-ol **[1]**. Collect the fraction that boils off around 82 °C **[1]**   = **[5]**

b Amount = 6.15 g ÷ 123 g mol^{-1} = 0.0500 mol **[1]** vol of NaOH = 0.0500 mol ÷ 2.0 mol dm^{-3} = 0.025 dm^3 = 25 cm^3 **[1]**   = **[2]**

c 100% yield would give 0.0500 mol of propan-2-ol = 0.0500 mol × 60 g mol^{-1} = 3.00 g **[1]**.
80% yield = 3.00 × 80/100 = 2.4 g **[1]**   = **[2]**

d One of: the reaction did not go to completion; it is an equilibrium reaction;
some propene was formed by elimination.   = **[1]**

Unit 4

Topic 4.1 Energetics II

1 a i

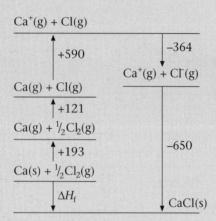

ii

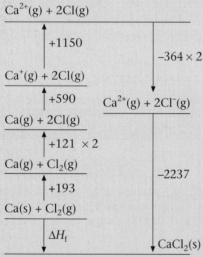

b i ΔH_f of $CaCl_2(s)$

$= +193 +590 +1150 + 2 \times 121 + 2 \times (-364) + (-2237)$

$= -790 \text{ kJ mol}^{-1}$

ii ΔH_f of $CaCl(s)$

$= +193 +590 +121 + (-364) + (-650) = -110 \text{ kJ mol}^{-1}$

iii ΔH_r of $(2CaCl(s) \rightarrow CaCl_2(s) + Ca(s))$

$= -790 -2 \times (-110) = -570 \text{ kJ mol}^{-1}$

Therefore products are more thermodynamically stable than reactants. Thus CaCl will disproportionate.

2 NaF: Experimental lattice energy $\cong$ theoretical lattice energy.

Therefore it is nearly 100% ionic.

MgI_2: Experimental >> theoretical.

Therefore it is considerably covalent, because the doubly charged Mg^{2+} ion is much more polarising than the singly charged Na^+ ion.

Also the much bigger I^- ion is more polarisable than the smaller F^- ion.

Topic 4.2 The Periodic Table II

1 a i $NaCl(s) + aq \rightarrow Na^+(aq) + Cl^-(aq)$

ii $SiCl_4 + 2H_2O \rightarrow SiO_2 + 4HCl$

iii $PCl_5 + 4H_2O \rightarrow H_3PO_4 + 5HCl$.

b NaCl is ionic and so just dissolves

$SiCl_4$ and PCl_5 are covalent and so react to give HCl.

c A lone pair of electrons on the oxygen atom of H_2O forms a dative bond into the empty 3d orbital in the silicon atom. This releases enough energy to break the Si–Cl bond.

In CCl_4, the carbon atom has no 2d orbital and so a dative bond cannot form before the C–Cl bond breaks, and the four large chlorine atoms around the small carbon atom sterically hinder the approach of a water molecule.

2 As a base:

$Al(OH)_3(s) + 3H^+(aq) \rightarrow Al^{3+}(aq) + 3H_2O(l)$

and as an acid:

$Al(OH)_3(s) + 3OH^-(aq) \rightarrow [Al(OH)_6]^{3-}(aq)$

3 a i $Mg(OH)_2(s) + 2H^+(aq) \rightarrow Mg^{2+}(aq) + 2H_2O(l)$
ii $PbO(s) + 2H^+(aq) \rightarrow Pb^{2+}(aq) + H_2O(l)$
iii SO_2 + acid: no reaction.
b i $Mg(OH)_2$ + alkali: no reaction
ii $PbO(s) + 2OH^-(aq) + H_2O(l) \rightarrow Pb(OH)_4^{2-}(aq)$
iii $SO_2(g) + 2OH^-(aq) \rightarrow SO_3^{2-}(aq) + H_2O(l)$

4 a $PbO_2 + 4HCl \rightarrow PbCl_2 + 2H_2O + Cl_2$
b $Sn^{2+}(aq) + 2Fe^{3+}(aq) \rightarrow Sn^{+}(aq) + 2Fe^{2+}(aq)$

Topic 4.3 Chemical equilibria II

1

	$2SO_3$	⇌	$2SO_2$ +	O_2
Start (moles)	0.0200		0	0
Change (moles)	−0.0058		+0.0058	+0.0029

> ### Examiner's comments
>
> 1 mole of SO_3 gives 1 mole of SO_2 and $^1/_2$ mole of O_2.

Equilib. moles	0.0142	0.0058	0.0029
$[\ ]_{eq}$/mol dm^{-3}	0.0142/1.52	0.0058/1.52	0.0029/1.52
	= 0.00934	= 0.00382	= 0.00191

$K_c = \dfrac{[SO_2]^2 \times [O_2]}{[SO_3]^2}$

$= \dfrac{(0.00382)^2 \times 0.00191}{(0.00934)^2}$

$= 3.19 \times 10^{-4}$ mol dm^{-3}

2

	$2NO(g)+$	$O_2(g)$	⇌	$2NO_2(g)$
Start (moles)	1.0	1.0		0
				Total = 2 mol
Change (moles)	−0.60	−0.30		+0.60
Equilib. (moles)	0.40	0.70		0.60
				Total = 1.7 mol
Equilib. mole fraction	0.40/1.7	0.70/1.7		0.60/1.7
	= 0.235	= 0.412		= 0.353
Partial pressure/atm	0.235 × 4	0.412 × 4		0.353 × 4
	= 0.94	= 1.65		= 1.41

$K_p = \dfrac{(p \text{ of } NO_2)^2}{(p \text{ of } NO)^2 \times p \text{ of } O_2}$

$= \dfrac{(1.41)^2}{(0.94)^2 \times 1.65}$

$= 1.4$ atm^{-1}

3 a $\dfrac{[SO_3]_{initial}^2}{[SO_2]_{initial}^2 \times [O_2]_{initial}}$

$= \dfrac{(0.2)^2}{(0.2)^2 \times 0.1}$

$= 10$ mol^{-1} dm^3

which is **not** equal to K_c, so it is **not** at equilibrium. As the quotient $<K_c$, the reaction moves left to right, until equilibrium is reached.

> ### Examiner's comments
>
> The [ ] term only equals K when the system is in equilibrium.

b i Because reaction L→R is **exo**thermic, a **decrease** in temperature causes K_c to increase, and so the position of equilibrium moves to the right.

ii As the reaction going from right to left causes an increase in the number of moles of gas, a decrease in pressure will cause the position of equilibrium to shift from right to left. K_c is unaltered.

iii There will be **no** effect on either. A catalyst speeds up the rate of both forward and back reactions **equally**, and so equilibrium is reached sooner.

Topic 4.4 Acid–base equilibria

1 $C_2H_5COOH \rightleftharpoons H^+ + C_2H_5COO^-$
$pK_a = 4.87$
Therefore $K_a = $ inverse $\log(-4.87) = 1.35 \times 10^{-5}$ mol dm^{-3}
$[H^+] = [C_2H_5COO^-]$
$K_a = \dfrac{[H^+] \times [C_2H_5COO^-]}{[C_2H_5COOH]}$
$= \dfrac{[H^+]^2}{[C_2H_5COOH]} = 1.35 \times 10^{-5}$ mol dm^{-3}
$[H^+] = \sqrt{(K_a \times [C_2H_5COOH])}$
$= \sqrt{(1.35 \times 10^{-5} \times 0.22)} = 1.72 \times 10^{-3}$ mol dm^{-3}
$pH = \sqrt{-\log[H^+]} = 2.76$

2 a End point is at 25 cm^3 NaOH.

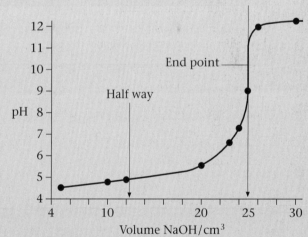

Half way to end point is 12.5 cm^3, which is when [HA] = [A$^-$].
pH at this point is 4.95.
Therefore $pK_a = 4.95$.

b Phenolphthalein.
3 a A buffer is a solution of known pH, which will

> ### Examiner's comments
>
> This is a suitable indicator because its range of pH for its entire colour change is within the vertical part of the graph.

resist change in pH when contaminated with small amounts of either acid or base. It consists of any weak acid and its conjugate base, such as ethanoic acid and sodium ethanoate.

Examiner's comments

Do not say that the pH is constant. 'Nearly constant' is acceptable.

b The salt is fully ionised:
$$CH_3COONa(aq) \rightarrow CH_3COO^-(aq) + Na^+(aq)$$
The weak acid is partially ionised:
$$CH_3COOH(aq) \rightleftharpoons H^+(aq) + CH_3COO^-(aq)$$
The CH_3COO^- ions from the salt suppress most of the ionisation of the acid.
If H^+ ions are added to the solution, almost all of them are removed by reaction with the **large reservoir** of CH_3COO^- ions from the salt, thus the pH hardly alters.
$$H^+(aq) + CH_3COO^-(aq) \rightarrow CH_3COOH(aq)$$
If OH^- ions are added, almost all of them are removed by reaction with the **large reservoir** of CH_3COOH molecules from the weak acid, thus the pH hardly alters.
$$OH^-(aq) + CH_3COOH \rightarrow CH_3COO^-(aq) + H_2O(l)$$

Examiner's comments

Four equations are necessary for a good answer, with one of them as an equilibrium reaction.

4

$[H^+] = K_a \times$ [weak acid]/[salt]
[weak acid] = 0.44 mol dm^{-3}
Amount of salt = 4.4/82 = 0.0537 mol.
Therefore [salt] = 0.0537/0.1 = 0.537 mol dm^{-3}
$[H^+] = \dfrac{1.74 \times 10^{-5} \times 0.44}{0.537}$
= 1.43×10^{-5} mol dm^{-3}
Therefore pH = $-\log(1.43 \times 10^{-5})$ = 4.85.

Examiner's comments

The concentration **not** the number of moles should be used in buffer calculations.

Topic 4.5 Organic chemistry II

1 a

(a)

b

(b)

Examiner's comments

If there is a double bond, look for geometric isomers. Draw them with the correct bond angles, not 90°. Look for four different groups around a carbon atom for optical isomerism.

2

$$CH_3CHICH_3 \xrightarrow{\text{Mg/dry ether}} CH_3CH(MgI)CH_3$$

$$CH_3CH(MgI)CH_3 \xrightarrow[\text{HCl(aq)}]{CO_2(s) \text{ then}} CH_3CH(CH_3)COOH$$

or
$$CH_3CHICH_3 \xrightarrow{KCN} CH_3CH(CH_3)CN$$

$$CH_3CH(CH_3)CN \xrightarrow[\text{with HCl(aq)}]{\text{heat under reflux}} CH_3CH(CH_3)COOH$$

Examiner's comments

An increase in the number of carbon atoms in a chain indicates the use of a Grignard reagent or a cyanide.

3 a Add water to each: ethanoyl chloride gives off steamy fumes of HCl; ethanoic acid shows no visible reaction.
Add PCl_5 to each: ethanoic acid gives off steamy fumes of HCl; ethanoyl chloride shows no visible reaction.

b Add to a solution of iodine and sodium hydroxide: propanone gives a pale yellow precipitate; propanal gives no precipitate.
Add ammoniacal silver nitrate solution and warm: propanal gives a silver mirror; propanone gives no mirror.

Answers to Practice Test Unit 4

The allocation of marks follows Edexcel mark schemes.
The marks that you will need for each grade are approximately:

A	70%
B	61%
C	52%
D	44%
E	36%

1. a

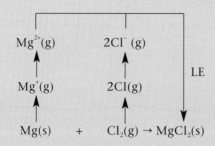

Correct cycle with all state symbols **[2]**
$\Delta H_{formation}$ = sum of all other changes
$-642 = +150 +736 + 1450 + 2 \times 121$
$+ 2 \times (-364) + LE$ **[2]**
$LE = -2492$ kJ mol^{-1} **[1]** = **[5]**

b i The ionic radius of Sr^{2+} is < that of Ba^{2+} **[1]**, and so there is a stronger force of attraction between cation and anion in $Sr(OH)_2$ than in $Ba(OH)_2$ **[1]** = **[2]**

ii Energy is released when the δ-oxygen atom in water **[1]** is attracted to the positive cation **[1]** = **[2]**

iii $\Delta H_{solution}$ = $-$Lattice energy + $\Delta H_{hydration}$ of cation + $\Delta H_{hydration}$ of the anions **[1]**
ΔH_{sol} $Ba(OH)_2$ = $+2228 - 1360 - 920 = -52$kJ **[1]**
ΔH_{sol} $Sr(OH)_2$ = $+2354 - 1480 - 920 = -46$kJ **[1]**
so $\Delta H_{solution}$ of $Ba(OH)_2$ is 6 kJ more exothermic and so it is more soluble **[1]** = **[4]**

Examiner's comments

a Remember that lattice energies are negative.
b iii If you forgot to multiply -460 by 2 (there are 2 OH$^-$ ions) you would lose 1 mark. Another way to score the 3rd and 4th marks is: from $Sr \rightarrow Ba$, the lattice energy decreases by 126 kJ **[1]**, but $\Delta H_{hydration}$ of the cation only decreases by 120 kJ **[1]**

---

2 a i $CH_3CH_2CH_2C\equiv N$ = **[1]**
ii $LiAlH_4$ or $NaBH_4$ = **[1]**
iii $CH_3CH_2CH_2CH_2NH_2$ = **[1]**
b i The lone pair of electrons on the N atom make it a base = **[1]**
ii $RNH_2 + H^+ \rightarrow RNH_3^+$ = **[1]**
c i

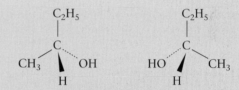

ii It forms a substituted amide = **[1]**

---

iii One of: the reaction is faster, or no catalyst is required, or with ethanol the reaction is an equilibrium = **[1]**

iv

H H
H O—C—C—H
H—C—C   H H
H   O

Examiner's comments

a As a full structural formula was not asked, you need not show all the atoms.
b Any amine would do.
c Full structural formulae were asked, so all atoms and bonds must be shown.

---

3 a

C₂H₅   C₂H₅
C   C
CH₃ OH   HO   CH₃
H   H

b Reagents: C_2H_5CHO**[1]** and CH_3MgBr **[1]** or CH_3CHO **[1]** and C_2H_5MgBr **[1]**
Conditions; dry ether solvent **[1]** = **[3]**
c i **Y** is $CH_3CH_2COCH_3$ = **[1]**
ii **Y** is a ketone = **[1]**
iii Add 2,4-dinitrophenylhydrazine **[1]** which gives an orange ppt. **[1]**
Add Fehling's solution (or ammoniacal silver nitrate) **[1]** which stays blue (does not give a silver mirror) **[1]** = **[4]**
iv CHI_3 **[1]** and C_2H_5COONa **[1]** = **[2]**

Examiner's comments

a You must make an attempt at a 3-D drawing and the two isomers must be mirror images of each other.
b iii You must show that it is a carbonyl compound and then that it is not an aldehyde.
b iv C_2H_5COOH would be accepted as the other organic product.

---

4 a $K_p = \dfrac{p(AlCl)^3}{p(AlCl_3)}$ = **[1]**

b Amount of AlCl = 0.63g/62.5 g mol^{-1} = 0.01 mol
Amount of $AlCl_3$ = 0.67 g/133.5 g mol^{-1} = 0.005mol **[1]** total moles = 0.015
mole fractions: AlCl = 0.667, $AlCl_3$ = 0.333 **[1]**
partial pressures: p(AlCl) = 0.667 × 2 = 1.33 atm
$p(AlCl_3)$ = 2 × 0.333 = 0.667 atm **[1]**
K_p = (1.33 atm)^3/0.667 atm = 3.56 atm^2 **[1]** = **[4]**

c An increase in temperature shifts the equilibrium in the endothermic direction. $\therefore \Delta H$ is positive = **[1]**

d K_p stays constant **[1]** The position of equilibrium moves to the side with fewer gas molecules **[1]**, which is to the left **[1]** = **[3]**

e Mix scrap aluminium metal with $AlCl_3$ and heat to form $AlCl$ **[1]**, remove from excess Al, then cool (or increase pressure) and pure Al and $AlCl_3$ are formed. The latter is then used again **[1]** = **[2]**

Examiner's comments

a Solids do not appear in K_p expressions.
b Partial pressure is mole fraction × total pressure.
c K is only altered by temperaure.

---

5 a i $MgO(s) + H_2SO_4(aq) \rightarrow MgSO_4(aq) + H_2O(l)$
Correct equation **[1]** state symbols **[1]** = **[2]**

ii A white solid **[1]** reacts to form a colourless solution **[1]** = **[2]**

b i $P_4O_{10} + 12NaOH \rightarrow 4Na_3PO_4 + 6H_2O$
or $P_2O_5 + 6NaOH \rightarrow 2Na_3PO_4 + 3H_2O$
All species correct **[1]**, balanced **[1]** = **[2]**

ii For having 2 equations, one with H^+ (or HCl) and the other with OH^- (or NaOH) **[1]** and:
$Al(OH)_3 + 3H^+ \rightarrow Al^{3+} + 3H_2O$ or
$Al(OH)_3 + 3HCl \rightarrow AlCl_3 + 3H_2O$ **[1]** plus
$Al(OH)_3 + 3OH^- \rightarrow Al(OH)_6^{3-}$ or
$Al(OH)_3 + 3NaOH \rightarrow Na_3Al(OH)_6$ **[1]** = **[3]**

c MgO is basic, $Al(OH)_3$ is amphoteric and P_4O_{10} is acidic **[1]** showing the metallic character of the element decreasing across the period **[1]** = **[2]**

d As a Group is descended the oxides get more basic **[1]**. Indium is in the same Group as amphoteric aluminium, so indium oxide will be basic **[1]** = **[2]**

Examiner's comments

Many students are very bad at writing equations. You must make sure that you learn the equations in Topic 4.2 of the specification.
a Whenever you are asked to state what you would see, you must describe the appearance before **and** after reaction.

---

6 a An acid is a proton donor **[1]**,
Weak means that it is incompletely dissociated (ionised) **[1]**, but dilute means that its concentration is small **[1]** = **[3]**

b i $pH = -\log_{10}[H^+]$ = **[1]**

ii $K_a = 3.72 \times 10^{-8} = \dfrac{[H^+][OCl^-]}{[HOCl]}$ **[1]**

$[H^+] = 10^{-pH} = 5.89 \times 10^{-5} \text{ mol dm}^{-1}$ **[1]**

$[H^+] = [OCl^-]$ **[1]**

$\therefore [HOCl] = (5.89 \times 10^{-5})^2 \div 3.72 \times 10^{-8}$

$= 0.0932 \text{ mol dm}^{-3}$ **[1]** = **[4]**

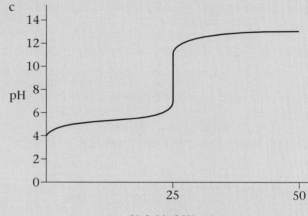

Vol. NaOH/cm³

Shape of curve correct **[1]**, start at pH = 4 **[1]**, vertical at 25 cm³ **[1]** and at a pH of between 7 and 11 **[1]**, curve flattening off at pH 13 **[1]** = **[5]**

d i $H_2SO_4 \rightarrow H^+ + HSO_4^-$ **[1]**
$HSO_4^- \rightleftharpoons H^+ + SO_4^{2-}$ **[1]** = **[2]**

ii The second ionisation is incomplete and is suppressed by the H^+ from the first ionisation, and so produces very few H^+ ions = **[1]**

Examiner's comments

a Don't confuse *weak* (how ionised) with *dilute* (how many moles).
b To get from pH to $[H^+]$, use the 10^x button on your calculator.
d Sulphuric acid is only strong in its 1st ionisation, so this solution has a pH of 0.98 not 0.699.

Unit 5

Topic 5.1 Redox equilibria

1 a i $Cr_2O_7^{2-}(aq) + 14H^+(aq) + 6e^- \rightarrow 2Cr^{3+}(aq) + 7H_2O$

ii $Sn^{4+}(aq) + 2e^- \rightarrow Sn^{2+}(aq)$

iii $IO_3^-(aq) + 6H^+(aq) + 5e^- \rightarrow \frac{1}{2}I_2 + 3H_2O$

iv $I_2 + 2e^- \rightarrow 2I^-(aq)$

Examiner's comments

All are reductions and so have electrons on the left. The number of electrons equals the total change in oxidation number.

b i $Cr_2O_7^{2-}(aq) + 14H^+(aq) + 3Sn^{2+}(aq) \rightarrow 2Cr^{3+}(aq) + 7H_2O + 3Sn^{4+}(aq)$
$E_{reaction} = +1.33 - 0.15 = +1.18$ V.
Feasible as $E_{reaction} > 0$.

Examiner's comments

Equation **a ii** is multiplied by 3, reversed and added to equation **a i**.

ii $IO_3^-(aq) + 6H^+(aq) + 5I^-(aq) \rightarrow 3I_2 + 3H_2O$
$E_{reaction} = +1.19 - 0.54 = +0.65$ V.
Feasible as $E_{reaction} > 0$.

Examiner's comments

Equation **a iv** is multiplied by 5/2, reversed and added to equation **a iii**.

2 Equations are:
$Fe(s) + 2H^+(aq) \rightarrow Fe^{2+}(aq) + H_2(g)$
$5Fe^{2+}(aq) + MnO_4^-(aq) + 8H^+(aq) \rightarrow 5Fe^{3+}(aq) + Mn^{2+}(aq) + 4H_2O$
Amount of $MnO_4^- = 0.0235 \times 0.0200$
$= 4.70 \times 10^{-4}$ mol
Amount of Fe^{2+} in 25 cm³ sample = $4.70 \times 10^{-4} \times 5/1$
$= 0.00235$ mol

Examiner's comments

5/1 because there are $5Fe^{2+}$ to $1MnO_4^-$ in the equation.

Amount of Fe^{2+} in 250 cm³ sample = 0.0235 mol
Mass of Fe^{2+} = mass of Fe = $0.0235 \times 56 = 1.316$ g
Purity of iron in steel = $\dfrac{1.316 \times 100}{1.32} = 99.7\%$

3 Anode area: $Fe(s) \rightarrow Fe^{2+}(aq) + 2e^-$

Examiner's comments

<u>O</u>xidation at the <u>A</u>node, so electrons on the right.

Cathode area: $\frac{1}{2}O_2(aq) + H_2O + 2e^- \rightarrow 2OH^-(aq)$

Examiner's comments

<u>R</u>eduction at the <u>C</u>athode, so electrons on the left.

Between anode and cathode:
$2Fe(OH)_2(s) + 2OH^-(aq) \rightarrow Fe_2O_3(s) + 3H_2O(l) + 2e^-$

$\frac{1}{2}O_2(aq) + H_2O(l) + 2e^- \rightarrow 2OH^-(aq)$

Topic 5.2 Transition metal chemistry

1 The water ligands split the d orbitals into two of higher energy and three of lower energy. When white light shines on the solution, an electron absorbs visible light energy and moves (jumps) from the lower to the higher energy level, absorbing some of the red/green light and leaving blue light.

2 a $[Cr(H_2O)_6]^{3+}(aq) + 3OH^-(aq) \rightarrow Cr(OH)_3(s) + 6H_2O$

Examiner's comments

This reaction is deprotonation.

then $Cr(OH)_3(s) + 3OH^-(aq) \rightarrow Cr(OH)_6^{3-}(aq)$

Examiner's comments

This occurs because Cr(III) is amphoteric.

b $[Fe(H_2O)_6]^{2+}(aq) + 2OH^-(aq) \rightarrow Fe(OH)_2(s) + 6H_2O$
then no further reaction.
c $[Zn(H_2O)_4]^{2+}(aq) + 2OH^-(aq) \rightarrow Zn(OH)_2(s) + 4H_2O$
then $Zn(OH)_2(s) + 2OH^-(aq) \rightarrow Zn(OH)_4^{2-}(aq)$

3 a $[Fe(H_2O)_6]^{3+}(aq) + 3NH_3(aq) \rightarrow Fe(OH)_3(s) + 3NH_4^+(aq) + 3H_2O$

Examiner's comments

This reaction is deprotonation.

then no further reaction.
b $[Cu(H_2O)_6]^{2+}(aq) + 2NH_3(aq) \rightarrow Cu(OH)_2(s) + 2NH_4^+(aq) + 4H_2O$

Examiner's comments

This reaction is also deprotonation.

then $Cu(OH)_2(s) + 4NH_3(aq) + 2H_2O(l) \rightarrow [Cu(NH_3)_4(H_2O)_2]^{2+}(aq) + 2OH^-(aq)$

Examiner's comments

The overall reaction is ligand exchange.

4 a $VO_2^+(aq) + 4H^+(aq) + 3e^- \rightarrow V^{2+}(aq) + 2H_2O$

Examiner's comments

Vanadium(V) is reduced, so electrons are on the left. The oxidation number changes by 3, so there must be three electrons.

b $VO_2^+(aq) + 2H^+(aq) + e^- \rightarrow VO^{2+}(aq) + H_2O$

c $V^{3+}(aq) + H_2O \rightarrow VO^{2+}(aq) + 2H^+(aq) + e^-$

d $VO^{2+}(aq) + H_2O \rightarrow VO_2^+(aq) + 2H^+(aq) + e^-$

Topic 5.3 Organic chemistry III

1 a The bromine reacts with the iron catalyst to form $FeBr_3$:

$$2Fe + 3Br_2 \rightarrow 2FeBr_3$$

The $FeBr_3$ then reacts with more bromine to form Br^+ (the electrophile) and $FeBr_4^-$.

The intermediate cation loses H^+ to the $FeBr_4^-$.

Examiner's comments

Check that:

● the arrow starts on the delocalised ring and goes towards the Br^+ (and not to the + of Br^+)

● the intermediate has a broken delocalised ring with a + inside it

● the arrow starts from the σ bond of the ring/H atom and goes inside the hexagon (but not directly to the +).

b It is energetically favourable for the intermediate cation to lose an H^+ and gain the stability of the benzene ring, rather than add Br^- as happens with alkenes.

2 The first step is the addition of a CN^- ion. HCN is too weak an acid to produce a significant amount of CN^- ions. NaOH will deprotonate HCN molecules, producing the necessary CN^- ions.

3 a If the temperature is too low (<5 °C), the rate is too slow.

If the temperature is too high (>5 °C), the benzene diazonium chloride will decompose.

b $C_6H_5N=NC_6H_4O^-$ but $C_6H_5N=NC_6H_4OH$ is acceptable.

Topic 5.4 Chemical kinetics II

1 Rate = $k[HI]^2$

$$k = \frac{\text{rate}}{[HI]^2}$$

$$= \frac{2.0 \times 10^{-4} \text{ mol dm}^{-3} \text{ s}^{-1}}{(0.050 \text{ mol dm}^{-3})^2}$$

$$= 0.080 \text{ s}^{-1} \text{ mol}^{-1} \text{ dm}^3$$

2 Carry out the following procedure:

● Place equal volumes of solution, e.g. 50 cm^3 of 0.10 mol dm^{-3} ethanoic acid and methanol in flasks in a thermostatically controlled tank at 60 °C.

● Mix, start the clock and replace in the tank.

● At intervals of time, pipette out 10 cm^3 portions and add to 25 cm^3 iced water in a conical flask.

● Rapidly titrate with standard sodium hydroxide solution using phenolphthalein as the indicator.

● Repeat several times.

● Plot a graph of the titre (which is proportional to the amount of ethanoic acid left) against time.

3 Time taken for the concentration to halve from 1.6 to 0.8 mol dm^{-3} = 26 minutes.

Time taken to halve again to 0.4 mol dm^{-3} = 26 minutes.

Time taken to halve again to 0.2 mol dm^{-3} = 26 minutes.

$t_{1/2}$ is a constant, therefore reaction is 1st order.

Topic 5.5 Organic chemistry IV

1 a Add 2,4-dinitrophenylhydrazine. Both give an orange precipitate.

Add ammoniacal silver nitrate solution. Only pentanal will give a silver mirror on warming.

Examiner's comments

Fehling's solution, which gives a red precipitate, could be used in place of ammoniacal silver nitrate.

b Heat under reflux with aqueous sodium hydroxide.

Examiner's comments

Halogenoalkanes are covalent and so must first be hydrolysed to produce halide ions.

Cool and acidify with dilute nitric acid, then add silver nitrate solution.

The 2-bromo compound gives a cream precipitate insoluble in dilute ammonia but soluble in concentrated ammonia.
The 2-chloro compound gives a white precipitate which dissolves in dilute ammonia.

c   Warm with dilute sulphuric acid and potassium dichromate(VI) solution and distill off any product into ammoniacal silver nitrate solution.
Methylpropan-2-ol does not change the colour of the potassium dichromate(VI).
Methylpropan-1-ol turns it from orange to green and the distillate gives a silver mirror.
Butan-2-ol turns it from orange to green and the distillate has no effect on the silver nitrate solution. To confirm, add a few drops of the butan-2-ol to iodine and aqueous sodium hydroxide and warm gently. A yellow precipitate of iodoform will be produced.

Examiner's comments

The 1° alcohol is partially oxidised to an aldehyde.
The 2° alcohol is oxidised to a ketone.
The 3° alcohol is not oxidised.
Butan-2-ol contains the $CH_3CH(OH)$ group and so gives a positive iodoform test.

2   The substance is probably aromatic (≥ 6 carbon atoms and about the same number of hydrogen atoms).
120 is the molecular ion.
105 is 15 less than 120 and is probably caused by loss of CH_3.
77 is probably caused by the $(C_6H_5)^+$ group (77 is 43 less than 120 and is probably caused by loss of $COCH_3$).
X is probably $C_6H_5COCH_3$.
The 120 peak is caused by $(C_6H_5COCH_3)^+$,
the 105 peak by $(C_6H_5CO)^+$,
and the 77 peak by $(C_6H_5)^+$.

3   3200 cm^{-1} is due to O–H, 1720 cm^{-1} is due to C=O, and 1150 cm^{-1} is due to C–O.

4   Six of the carbon atoms are in a benzene ring. The remainder are either C_2H_5 group or $2 \times CH_3$ groups.

Y is $C_6H_5CH_2CH_3$.

[5 H's in C_6H_5, 2 H's in CH_2 and 3 H's in CH_3.]

Z is $C_6H_4(CH_3)_2$.

[6 H's in $2 \times CH_3$ groups, 4 H's in C_6H_4.]

Answers to Practice Test Unit 5

The allocation of marks follows Edexcel mark schemes.
The marks that you will need for each grade are

approximately:

A	68%
B	59%
C	50%
D	41%
E	32%

1 a i   Consider experiments 1 & 2: when $[RCH_2Cl]$ is increased 3 times, rate also increases 3 times [1], ∴ 1st order with respect to $[RCH_2Cl]$. [1]
Consider experiments 1 & 3: when concentrations of both are doubled, the rate is increased 4 times [1], ∴1st order with respect to $[OH^-]$ as well [1]   = [4]
   ii   Rate = k $[RCH_2Cl] \times [OH^-]$ = [1]
   iii   k = rate/($[RCH_2Cl] \times [OH^-]$)
   = $\dfrac{4.0 \times 10^{-4} mol\ dm^{-3}\ s^{-1}}{0.050\ mol\ dm^{-3} \times 0.10\ mol\ dm^{-3}}$
   = 0.080 [1] $mol^{-1}\ dm^3\ s^{-1}$ [1]   = [2]
   iv   As it is 1st order in both, it is a S_n2 mechanism [1].

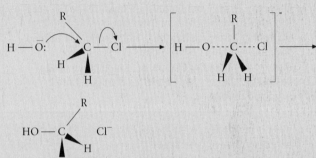

Curly arrows from O of OH^- to C and from C–Cl σ bond to Cl [1], correct transition state [1]   = [3]

Examiner's comments

i   You must make it clear what data you are using and how you arrive at each order.
iv   Make sure that all your curly arrows start either at a bond and go to an atom or at an atom and go to a bond.

2 a   The electrophile is CH_3CO^+ [1]. The equation is:
$CH_3COCl + AlCl_3 \rightarrow CH_3CO^+ + AlCl_4^-$ [1]   = [2]

b

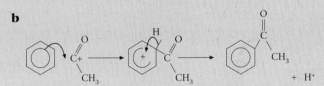

Curly arrow from ring to C of CO^+ [1], correct intermediate [1], curly arrow from C–H σ bond to ring [1]   = [3]
c   React the phenylethanone, $C_6H_5COCH_3$, with HCN [1], with a trace of base (or in a solution buffered anywhere between pH 5 to 9 or add a mixture of HCN and KCN) [1]. This produces

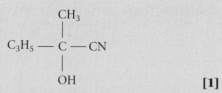

$$[1]$$

Now add aqueous sulphuric acid (or any named acid) **[1]** and heat under reflux **[1]**
= **[5]**

d Mix known amounts of ester and OH⁻ **[1]**
remove sample **[1]**
at known time **[1]**
quench the reaction by adding ice cold water **[1]**
titrate unreacted OH⁻ with acid **[1]**
Calculate rate = change in [OH⁻] ÷ time **[1]**
Repeat with double [ester] and the same [OH⁻] **[1]**
Repeat with double [OH⁻] and the same[ester] **[1]**
= **[8]**

Examiner's comments

b The curly arrow must go towards the C of the CO **not** the C of the CH₃ group. The intermediate must have a broken ring going across all but the C atom which has formed the bond with the CH₃CO group.

d The pH would hardly change, so a method involving measuring the pH change over time would score a maximum of 6.

3 a The reagent for step 1 is magnesium **[1]**, and the conditions are dry ether **[1]** = **[2]**

b i Potassium dichromate(VI) **[1]** and sulphuric acid **[1]** = **[2]**

ii Distil off **D** as it is formed **[1]** = **[1]**

iii Ethanoic acid (CH₃COOH) or ethanoate ions (CH₃COO⁻) **[1]** = **[1]**

c **E** is CH₃CH₂CH(OH)CH₃ **[1]** for any secondary alcohol and **[1]** for the correct formula = **[2]**

d i **F** is a carbonyl compound as it reacts with 2,4-dinitrophenyl hydrazine **[1]**, but it must be a ketone and not an aldehyde as it does not react with ammoniacal silver nitrate **[1]**
= **[2]**

ii In order to do the iodoform reaction **F** must have the CH₃CO group **[1]** and the products are CHI₃ **[1]** and CH₃CH₂COO⁻ **[1]** = **[3]**

iii **F** is CH₃CH₂COCH₃ **[2]** = **[2]**

e The species are: (CH₃CH₂COCH₃)⁺ for m/e = 72 **[1]**, (COCH₃)⁺ for 43 **[1]** and (CH₃CH₂)⁺ for 29 **[1]**
= **[3]**

Examiner's comments

b iii The aldehyde **D** is oxidised to an acid by the silver/ammonia complex ions.

c Grignard reagents react with an aldehyde to give a secondary alcohol.

d ii The only substances to do the iodoform reaction are alcohols with a CH₃CH(OH) group, ketones with a CH₃CO group and ethanal.

e Don't forget the charge on the species in a mass spectrum.

4 a Dip a platinum electrode **[1]** into a solution which is 1 mol dm⁻³ in both Fe²⁺ and Fe³⁺ ions **[1]**. This is connected via a salt bridge (containing potassium chloride solution) **[1]** to a standard hydrogen electrode **[1]**. The potential is measured with a high resistance voltmeter (or a potentiometer) **[1]** = **[5]**

b i Au³⁺ + 3e⁻ → Au **[1]**
Fe²⁺ → Fe³⁺ + e⁻ **[1]** = **[2]**

ii Au³⁺ + 3Fe²⁺ → Au + 3Fe³⁺ = **[1]**

iii Au³⁺ + 3e⁻ → Au $\qquad$ $E°$ (gold)
3Fe²⁺ → 3Fe³⁺ + 3e⁻ $\qquad$ $E°$ = −0.77 V
adding the two half-equations gives $E°$ cell
$E°$ (gold) + (−0.77) = $E°$ (cell) = + 0.73 **[1]**
$E°$ (gold) = + 0.73 + 0.77 = + 1.50V **[1]** = **[2]**

Examiner's comments

b i Reduction (gain of electrons) takes place at the gold electrode because it acts as the cathode. Thus Au³⁺ ions gain electrons that flowed in the external circuit from Fe²⁺ ions which lost them.

iii Multiplying the Fe²⁺ half-equation by 3 does not alter its $E°$ value.
Another way of doing the calculation is to use the standard reduction potentials thus:
$E°$ of oxidising agent (gold) − $E°$ of reducing agent (iron) = $E°$ of the cell.

5 a

		3d	4s
Fe :	[Ar]	↑↓ ↑ ↑ ↑ ↑	↑↓
Cr :	[Ar]	↑ ↑ ↑ ↑ ↑	↑
Cr ³⁺:	[Ar]	↑ ↑ ↑ ☐ ☐	☐

b FeCrO₄ + 4C → Fe + Cr + 4CO
species **[1]**, balancing of equation **[1]** = **[2]**

c i [Fe(H₂O)₆]²⁺ = **[1]**

ii [Fe(H₂O)₆]²⁺ + 2OH⁻ → Fe(OH)₂ + 6H₂O
or [Fe(H₂O)₆]²⁺ + 2OH⁻ → [Fe(OH)₂(H₂O)₄] + 2H₂O
correct iron species in product **[1]** balance **[1]** = **[2]**

iii The precipitate's colour is pale green = **[1]**

iv [Cr(OH)₆]³⁻ = **[1]**

v The solution of **R** would first give a green precipitate **[1]**, which forms a green solution with excess acid **[1]** = **[2]**

d i The standard electrode potential is the potential difference between a standard hydrogen electrode and the half-cell **[1]** where all concentrations are 1 mol dm⁻³ and the pressure of all gases is 1 atm. **[1]** = **[2]**

ii All four halogens will oxidise Cr²⁺ = **[1]**

iii Both bromine and iodine will not oxidise Cr³⁺ further = **[1]**

iv Mix the blue Cr^{2+} solution separately with bromine and iodine **[1]**, the solution goes green and stays green with excess halogen **[1]** = **[2]**

Answers to Practice Test Unit 6B

The allocation of marks follow Edexcel mark schemes. The marks that you will need for each grade are approximately:

A	66%
B	57%
C	48%
D	40%
E	32%

1 a Amount of NaOH = 0.100 mol dm^{-3} × 0.0280 dm^3
= 0.00280 mol **[1]** = moles of H$^+$
Ratio H$^+$ to NH$_4^+$ = 1 : 1.
Amount of NH$_4^+$ in 25 cm^3 = 0.00280 mol.
Amount in 250 cm^3 = 0.0280 mol **[1]**
Amount of $(NH_4)_2SO_4$ = $\frac{1}{2}$ × 0.0280 = 0.0140 mol **[1]**,
mass of $(NH_4)_2SO_4$ = 0.0140 mol × 132 g mol^{-1} = 1.848 g **[1]**,
% $(NH_4)_2SO_4$ in fertiliser = 1.848 × $100/3.80$ = 48.6% **[1]** = **[5]**

b Any two of: not all ammonia driven off **[1]**; ammonia incompletely absorbed by the HCl **[1]**; some ammonia gas escapes **[1]** = **[2]**

c i $Ba^{2+}(aq) + SO_4^{2-}(aq) \rightarrow BaSO_4(s)$
Species **[1]**, state symbols **[1]** = **[2]**

ii To ensure that all the sulphate ions were precipitated = **[1]**

d Many carbonates will decompose on heating **[1]**, and the CO_2 gas evolved will result in a lower mass being recorded **[1]**. An example is $CaCO_3$ **[1]**
$CaCO_3 \rightarrow CaO + CO_2$ **[1]** = **[4]**

Answer any two of the section B questions

2 a Any two of : it is considerably cheaper **[1]**; less risk of leaching **[1]** it does not effect the pH of the soil **[1]** it releases the nitrogen slowly **[1]** = **[2]**

b The smaller the K_a of the conjugate acid, the stronger the base **[1]**. Therefore ammonia is the stronger base **[1]** = **[2]**

c Polymerisation occurs when many molecules join to form a long chain **[1]**; condensation is with the elimination of water or a small inorganic molecule such as HCl **[1]**.
A polyamide contains the –CONH– link **[1]** as in:

[1]

d i The reagents are : liquid bromine **[1]** and concentrated sodium hydroxide **[1]**. The condition is that it must be heated **[1]** = **[3]**

ii NH_2NH_2 = **[1]**

e i Solids are not included in K_p expressions and ammonium nitrate is a solid. = **[1]**

ii As the reaction is endothermic, the energy level of the products is higher than that of the reactants **[1]**, therefore the reaction is said to

be thermodynamically stable **[1]** and the equilibrium lies to the left **[1]**.
As the reaction proceeds on moderate heating, the activation energy is fairly small **[1]** and so the reaction is kinetically unstable **[1]**

3 a i The reagents are potassium dichromate(VI) **[1]** and sulphuric acid **[1]**. The conditions are careful heating and distil off the aldehyde as it is formed **[1]**.   = **[3]**

ii The conditions for the addition of HCN to a carbonyl compound are KCN in aqueous ethanol at a pH of 8 (or a mixture of KCN and HCN) **[1]**
The mechanism is:

[1] for a curly arrow going from the C in CN^-
[1] for a curly arrow going from the $C=O$ π bond to the oxygen atom
[1] for the intermediate with its – charge   = **[4]**

iii $CH_3CH_2CH_2CH(CH_3)CH(OH)CN + HCl + 2H_2O$
$\rightarrow CH_3CH_2CH_2CH(CH_3)CH(OH)COOH$
$+ NH_4Cl$ **[1]**,
correct formula of organic product **[1]**   = **[2]**

iv

HA

= **[1]**

b i $K_a = \dfrac{[H_3O^+] \times [A^-]}{[HA]}$ or $[H^+]$ instead of $[H_3O^+]$ **[1]**

$[H_3O^+] = [A^-] = \sqrt{(K_a\,c)} = \sqrt{(4.5 \times 10^{-3} \times 0.0100)}$
$= 6.71 \times 10^{-4}$ mol dm^{-3} **[1]**
pH $= -\log(6.71 \times 10^{-4}) = 3.17$ **[1]**   = **[3]**

ii The acid is partially ionised: $HA \rightleftharpoons H^+ + A^-$
The salt is totally ionised: $NaA \rightarrow Na^+ + A^-$ **[1]**
This suppresses the ionisation of the acid and so both $[HA]$ and $[A^-]$ are large compared to any H^+ or OH^- that may be added. **[1]**
When H^+ is added, it is removed by the reaction:
$H^+ + A^- \rightarrow HA$ **[1]**
If OH^- is added, it is removed by:
$OH^- + HA \rightarrow A^- + H_2O$ **[1]**   = **[4]**

4 a The Fe ion is 3+ whereas the Na ion is only 1+ **[1]**
The Fe^{3+} ion is much more polarising than the Na^+ ion **[1]**, and so it draws the electrons **[1]** from the large Cl^- ion towards itself and the bond becomes covalent. **[1]**   = **[4]**

b i $[Fe(H_2O)_6]^{3+} + H_2O \rightarrow H_3O^+ + [Fe(H_2O)_5OH]^{2+}$ **[1]**
acid $[Fe(H_2O)_6]^{3+}$: conjugate base $[Fe(H_2O)_5OH]^{2+}$ **[1]**
base H_2O: conjugate acid H_3O^+ **[1]**   = **[3]**

ii a red precipitate will form **[1]**
which stays with excess NaOH **[1]**
$[Fe(H_2O)_6]^{3+} + 3OH^- \rightarrow Fe(OH)_3 + 6H_2O$ **[1]** or
$[Fe(H_2O)_6]^{3+} + 3OH^- \rightarrow [Fe(H_2O)_3(OH)_3] + 3H_2O$
The reaction is deprotonation. **[1]**   = **[4]**

iii The bonding is dative covalent (coordinate) **[1]**
The ligands split the d-orbitals into two levels **[1]**
The light is absorbed and a d electron promoted from the lower to the higher level. **[1]**   = **[3]**

c Each oxygen is –2, $\therefore$ 4 oxygens = –8. The ion is 2–, and so the Fe is +6   (6+ and 8– make 2–) **[1]**
There are two ways. The first is: add excess acidified potassium iodide solution to an aliquot of the FeO_4^{2-} solution **[1]**, then titrate the liberated iodine against standard sodium thiosulphate solution **[1]**, adding starch when the solution becomes pale yellow and stopping when it is colourlesss **[1]**
The second method is: add excess acid to an aliquot **[1]**, then titrate against standard $FeSO_4$ solution **[1]** until the solution becomes very pale pink **[1]**   = **[4]**

Index

Appendix

 ## Organic reactions

1. AS and A2 (synoptic)

Alkanes, e.g. ethane CH_3CH_3

- Ethane → carbon dioxide and water
  - Reactant: oxygen (air)
  - Equation: $2CH_3CH_3 + 7O_2 \rightarrow 4CO_2 + 6H_2O$
  - Conditions: burn / spark
  - Classification: combustion
- Ethane → chloroethane
  - Reactant: chlorine
  - Equation: $CH_3CH_3 + Cl_2 \rightarrow CH_3CH_2Cl + HCl$
  - Conditions: sunlight
  - Classification: free radical substitution

Alkenes, e.g. ethene, $H_2C{=}CH_2$

- Ethene → ethane
  - Reactant: hydrogen
  - Equation: $H_2C{=}CH_2 + H_2 \rightarrow CH_3CH_3$
  - Conditions: heated nickel (or platinum) catalyst
  - Classification: addition or reduction or hydrogenation
- Ethene → 1,2-dibromoethane
  - Reactant: bromine
  - Equation: $H_2C{=}CH_2 + Br_2 \rightarrow CH_2BrCH_2Br$
  - Conditions: bubble ethene into bromine dissolved in hexane
  - Classification: electrophilic addition
- Ethene → bromoethane
  - Reactant: hydrogen bromide
  - Equation: $H_2C{=}CH_2 + HBr \rightarrow CH_3CH_2Br$
  - Conditions: mix gases at room temperature
  - Classification: electrophilic addition
- Ethene → ethan-1,2-diol
  - Reactant: potassium manganate(VII) solution
  - Equation: $H_2C{=}CH_2 + [O] + H_2O \rightarrow CH_2(OH)CH_2OH$
  - Conditions: a solution made alkaline with sodium hydroxide
- Ethene → poly(ethene)
  - Reactant: ethene
  - Equation: $n\,H_2C{=}CH_2 \rightarrow {\displaystyle +}(CH_2{-}CH_2{\displaystyle +})_n$
  - Conditions: 2000 atm pressure, 250 °C
  - Classification: addition polymerisation

Halogenoalkanes, e.g. 1-bromopropane

- 1-bromopropane → propan-1-ol
  - Reactant: sodium (or potassium) hydroxide
  - Equation: $CH_3CH_2CH_2Br + NaOH \rightarrow CH_3CH_2CH_2OH$
  - Conditions: heat under reflux in **aqueous** solution
  - Classification: nucleophilic substitution
- 1-bromopropane → propene
  - Reactant: potassium hydroxide
  - Equation: $CH_3CH_2CH_2Br + KOH \rightarrow CH_3CH{=}CH_2 + KBr + H_2O$
  - Conditions: heat under reflux in **ethanolic** solution
  - Classification: elimination
- 1-bromopropane → butanenitrile
  - Reactant: potassium cyanide
  - Equation: $CH_3CH_2CH_2Br + KCN \rightarrow CH_3CH_2CH_2CN + KBr$
  - Conditions: heat under reflux in a solution of ethanol and water
  - Classification: nucleophilic substitution
- 1-bromopropane → 1-aminopropane
  - Reactant: ammonia
  - Equation: $CH_3CH_2CH_2Br + 2NH_3 \rightarrow CH_3CH_2CH_2NH_2 + NH_4Br$
  - Conditions: heat a solution of ammonia in ethanol in a sealed tube
  - Classification: nucleophilic substitution

- 1-bromopropane → Grignard reagent (A2 only)
  - Reactant: Magnesium
  - Equation: $CH_3CH_2CH_2Br + Mg \rightarrow CH_3CH_2CH_2MgBr$
  - Conditions: warm (in water bath) under reflux in dry ether

Alcohols, e.g. ethanol
- ethanol → ethanal
  - Reactant: potassium dichromate(VI) + dilute sulphuric acid
  - Equation: $C_2H_5OH + [O] \rightarrow CH_3CHO + H_2O$
  - Conditions: heat carefully and distil out the aldehyde as it is formed
  - Classification: oxidation
- ethanol → ethanoic acid
  - Reactant: potassium dichromate(VI) + dilute sulphuric acid
  - Equation: $C_2H_5OH + 2[O] \rightarrow CH_3COOH + H_2O$
  - Conditions: heat under reflux
  - Classification: oxidation

NOTE: Secondary alcohols are oxidised to ketones and tertiary alcohols are not oxidised.

$CH_3CH(OH)CH_3 + [O] \rightarrow CH_3COCH_3 + H_2O$

$(CH_3)_3COH + [O] \rightarrow$ no reaction;   potassium dichromate solution stays orange

- ethanol → ethene
  - Reactant: concentrated sulphuric (or phosphoric) acid or aluminium oxide
  - Equation: $C_2H_5OH - H_2O \rightarrow H_2C{=}CH_2$
  - Conditions: heat
  - Classification: dehydration
- ethanol → chloroethane
  - Reactant: phosphorus pentachloride
  - Equation: $C_2H_5OH + PCl_5 \rightarrow C_2H_5Cl + POCl_3 + HCl$
  - Conditions: dry
- ethanol → bromoethane
  - Reactant: hydrogen bromide
  - Equation: $C_2H_5OH + HBr \rightarrow C_2H_5Br + H_2O$
  - Conditions: HBr made *in situ* from 50% sulphuric acid and solid potassium bromide
- ethanol → iodoethane
  - Reactant: hydrogen iodide
  - Equation: $C_2H_5OH + HI \rightarrow C_2H_5I + H_2O$
  - Conditions: HI made *in situ* from iodine and moist red phosphorus

2. A2 only

Grignard reagents, e.g. ethylmagnesium bromide
- Ethylmagnesium bromide → ethane
  - Reactant: water
  - Equation: $C_2H_5MgBr + H_2O \rightarrow C_2H_6 + Mg$ compounds
- Ethylmagnesium bromide → a secondary alcohol
  - Reactant: an aldehyde such as ethanal
  - Equation: $C_2H_5MgBr + CH_3CHO \rightarrow CH_3CH(OH)C_2H_5$
  - Conditions: dry ether solution, then hydrolyse with dilute acid
  - Classification: nucleophilic addition to the aldehyde
- Ethylmagnesium bromide → a tertiary alcohol
  - Reactant: a ketone such as propanone
  - Equation: $C_2H_5MgBr + CH_3COCH_3 \rightarrow (CH_3)_2C(OH)C_2H_5$
  - Conditions: dry ether solution, then hydrolyse with dilute acid
  - Classification: nucleophilic addition to the ketone
- Ethylmagnesium bromide → a carboxylic acid
  - Reactant: (solid) carbon dioxide
  - Equation: $C_2H_5MgBr + CO_2 \rightarrow C_2H_5COOH$
  - Conditions: dry ether solution, then hydrolyse with dilute acid

Carboxylic acids, e.g. ethanoic acid
- ethanoic acid → an ester
  - Reactant: ethanol
  - Equation: $CH_3COOH + C_2H_5OH {\rightleftharpoons} CH_3COOC_2H_5 + H_2O$
  - Conditions: heat under reflux with a few drops of concentrated sulphuric acid
  - Classification: esterification

- ethanoic acid → an alcohol
  Reactant: lithium aluminium hydride (lithium tetrahydridoaluminate(III))
  Equation: $CH_3COOH + 4[H] \rightarrow CH_3CH_2OH + H_2O$
  Conditions: dry ether solution then hydrolyse with dilute acid
  Classification: reduction
- ethanoic acid → an acid chloride
  Reactant: phosphorus pentachloride (or PCl_3 or $SOCl_2$)
  Equation: $CH_3COOH + PCl_5 \rightarrow CH_3COCl + POCl_3 + HCl$
  Conditions: dry
- ethanoic acid → a salt
  Reactant: sodium carbonate
  Equation: $2CH_3COOH(aq) + Na_2CO_3(s)$
  $$\rightarrow 2CH_3COONa(aq) + CO_2(g) + H_2O(l)$$
  Classification: neutralisation
- ethanoic acid → a salt
  Reactant: sodium hydrogencarbonate
  Equation: $CH_3COOH(aq) + NaHCO_3(s) \rightarrow CH_3COONa(aq) + CO_2(g) + H_2O(l)$
  Classification: neutralisation

Esters, e.g. ethyl ethanoate
- ethyl ethanoate → acid + alcohol
  Reactant: any aqueous strong acid such as dilute sulphuric acid
  Equation: $CH_3COOC_2H_5 + H_2O \rightleftharpoons CH_3COOH + C_2H_5OH$
  Conditions: heat under reflux
  Classification: reversible hydrolysis
- ethyl ethanoate → salt + alcohol
  Reactant: aqueous sodium hydroxide
  Equation: $CH_3COOC_2H_5 + NaOH \rightarrow CH_3COONa + C_2H_5OH$
  Conditions: heat under reflux
  Classification: hydrolysis (saponification)

Carbonyl compounds aldehydes, e.g. CH_3CHO, and ketones, e.g. CH_3COCH_3
- Both react with:
  Reactant: 2,4-dinitrophenylhydrazine
  Equation: $>C{=}O + NH_2NHC_6H_3(NO_2)_2 \rightarrow >C{=}N{-}NHC_6H_3(NO_2)_2 + H_2O$
  Conditions: mix solutions: orange precipitate observed
- Both react with:
  Reactant: hydrogen cyanide
  Equation: $>C{=}O + HCN \rightarrow >C(OH)CN$
  Conditions: potassium cyanide + some dilute sulphuric acid
  Classification: nucleophilic addition
- Both react with:
  Reactant: lithium aluminium hydride (or sodium borohydride)
  Equation: $CH_3CHO + 2[H] \rightarrow CH_3CH_2OH$ (a primary alcohol)
  $CH_3COCH_3 + 2[H] \rightarrow CH_3CH(OH)CH_3$ (a secondary alcohol)
  Conditions: dry ether, then hydrolyse with dilute acid
  Classification: reduction
- Aldehydes only react with:
  Reactant: Fehling's solution or ammoniacal silver nitrate
  Equation: $CH_3CHO + [O] + OH^- \rightarrow CH_3COO^- + H_2O$
  Conditions: warm
  Classification: oxidation
- Carbonyl compounds with a CH_3CO group give a yellow precipitate of iodoform with:
  Reactant: iodine and sodium hydroxide solution
  Equation: $CH_3COCH_3 + 3I_2 + 4OH^- \rightarrow CH_3COO^- + CHI_3 + 3I^- + 3H_2O$

Acid chlorides, e.g. ethanoyl chloride
- ethanoyl chloride → ethanoic acid
  Reactant: water
  Equation: $CH_3COCl + H_2O \rightarrow CH_3COOH + HCl$
  Classification: hydrolysis
- ethanoyl chloride → an ester
  Reactant: an alcohol
  Equation: $CH_3COCl + C_2H_5OH \rightarrow CH_3COOC_2H_5 + HCl$
  Conditions: rapid reaction at room temperature
  Classification: esterification

- ethanoyl chloride → ethanamide
  Reactant: ammonia
  Equation: $CH_3COCl + 2NH_3 \rightarrow CH_3CONH_2 + NH_4Cl$
- ethanoyl chloride → a substituted amide
  Reactant: an amine
  Equation: $CH_3COCl + C_2H_5NH_2 \rightarrow CH_3CONHC_2H_5 + HCl$

Amines, e.g. $C_2H_5NH_2$

- ethylamine → a salt
  Reactant: any acid such as hydrochloric
  Equation: $C_2H_5NH_2 + HCl \rightarrow C_2H_5NH_3^+Cl^-$
- ethylamine → a substituted amide
  Reactant: an acid chloride
  Equation: $C_2H_5NH_2 + CH_3COCl \rightarrow CH_3CONHC_2H_5 + HCl$

Nitriles, e.g. CH_3CN

- ethanenitrile → ethanoic acid
  Reactant: dilute sulphuric acid (or sodium hydroxide followed by acidification)
  Equation: $CH_3CN + H^+ + 2H_2O \rightarrow CH_3COOH + NH_4^+$
  Conditions: heat under reflux
  Classification: hydrolysis
- ethanenitrile → ethylamine
  Reactant: lithium aluminium hydride
  Equation: $CH_3CN + 4[H] \rightarrow CH_3CH_2NH_2$
  Conditions: dry ether then hydrolyse with dilute acid
  Classification: reduction

Amides, e.g. CH_3CONH_2

- ethanamide → methylamine
  Reactant: bromine and sodium hydroxide
  Equation: $CH_3CONH_2 + Br_2 + 2NaOH \rightarrow CH_3NH_2 + 2NaBr + H_2O + CO_2$
  Conditions: liquid bromine and conc sodium hydroxide
  Classification: Hofmann degradation reaction.
- ethanamide → ethanenitrile
  Reactant: phosphorus(V)oxide
  Equation: $CH_3CONH_2 - H_2O \rightarrow CH_3CN$
  Conditions: warm
  Classification: dehydration

Benzene

- benzene→ nitrobenzene
  Reactant: concentrated nitric acid
  Equation: $C_6H_6 + HNO_3 \rightarrow C_6H_5NO_2 + H_2O$
  Conditions: mix with concentrated sulphuric acid at 50 °C
  Classification: electrophilic substitution
- benzene → bromobenzene
  Reactant: bromine
  Equation: $C_6H_6 + Br_2 \rightarrow C_6H_5Br + HBr$
  Conditions: liquid bromine with an iron catalyst
  Classification: electrophilic substitution
- benzene → ethylbenzene
  Reactant: chloroethane
  Equation: $C_6H_6 + C_2H_5Cl \rightarrow C_6H_5C_2H_5 + HCl$
  Conditions: anhydrous aluminium chloride catalyst
  Classification: electrophilic substitution
- benzene → phenylethanone
  Reactant: ethanoyl chloride
  Equation: $C_6H_6 + CH_3COCl \rightarrow C_6H_5COCH_3 + HCl$
  Conditions: anhydrous aluminium chloride as catalyst
  Classification: electrophilic substitution

Alkylbenzenes, e.g ethylbenzene

- ethylbenzene → ethanoate ions
  Reactant: potassium manganate(VII) + sodium hydroxide
  Equation: $C_6H_5C_2H_5 + 6[O] + OH^- \rightarrow C_6H_5COO^- + 3H_2O + CO_2$
  Conditions: heat under reflux
  Classification: oxidation

Phenol

- phenol → sodium phenate
  - Reactant: sodium hydroxide
  - Equation: $C_6H_5OH + NaOH \rightarrow C_6H_5ONa + H_2O$
  - Classification: neutralisation
- phenol → 2,4,6-tribromophenol
  - Reactant: bromine
  - Equation: $C_6H_5OH + 3Br_2 \rightarrow HOC_6H_2Br_3 + 3HBr$
  - Conditions: aqueous; orange bromine water forms a white precipitate
  - Classification: electrophilic substitution
- phenol → phenyl ethanoate
  - Reactant: ethanoyl chloride
  - Equation: $C_6H_5OH + CH_3COCl \rightarrow CH_3COOC_6H_5 + HCl$
  - Classification: esterification

Nitrobenzene

Nitrobenzene → phenylamine
  - Reactant: tin and concentrated hydrochloric acid
  - Equation: $C_6H_5NO_2 + 6[H] \rightarrow C_6H_5NH_2 + 2H_2O$
  - Conditions: heat under reflux, then add sodium hydroxide
  - Classification: reduction

Phenylamine

Phenylamine → diazonium ion → azo dye
  - Reactant: step 1: nitrous acid at 5 °C; step 2: phenol
  - Equation: $C_6H_5NH_2 + 2H^+ + NO_2^- \rightarrow C_6H_5N_2^+ + 2H_2O$
  - $C_6H_5N_2^+ + C_6H_5OH \rightarrow HOC_6H_4-N{=}N-C_6H_5 + H_2O$
  - Conditions: mix phenylamine with sodium nitrite and hydrochloric acid at 5 °C, then add phenol in sodium hydroxide solution.

oxford
PRACTICE AIGUEL   540 PAR

The Periodic Table
of Elements

Group

Key

| Atomic number |
| Symbol |
| Name |
| Molar mass in g mol^{-1} |

Period	1	2												3	4	5	6	7	0
1	1 H Hydrogen 1																		2 He Helium 4
2	3 Li Lithium 7	4 Be Beryllium 9												5 B Boron 11	6 C Carbon 12	7 N Nitrogen 14	8 O Oxygen 16	9 F Fluorine 19	10 Ne Neon 20
3	11 Na Sodium 23	12 Mg Magnesium 24												13 Al Aluminium 27	14 Si Silicon 28	15 P Phosphorus 31	16 S Sulphur 32	17 Cl Chlorine 35.5	18 Ar Argon 40
4	19 K Potassium 39	20 Ca Calcium 40	21 Sc Scandium 45	22 Ti Titanium 48	23 V Vanadium 51	24 Cr Chromium 52	25 Mn Manganese 55	26 Fe Iron 56	27 Co Cobalt 59	28 Ni Nickel 59	29 Cu Copper 63.5	30 Zn Zinc 65.4		31 Ga Gallium 69.7	32 Ge Germanium 73	33 As Arsenic 75	34 Se Selenium 79	35 Br Bromine 80	36 Kr Krypton 84
5	37 Rb Rubidium 85	38 Sr Strontium 88	39 Y Yttrium 89	40 Zr Zirconium 91	41 Nb Niobium 93	42 Mo Molybdenum 96	43 Tc Technetium (99)	44 Ru Ruthenium 101	45 Rh Rhodium 103	46 Pd Palladium 106	47 Ag Silver 108	48 Cd Cadmium 112		49 In Indium 115	50 Sn Tin 119	51 Sb Antimony 122	52 Te Tellurium 128	53 I Iodine 127	54 Xe Xenon 131
6	55 Cs Caesium 133	56 Ba Barium 137	57 La▲ Lanthanum 139	72 Hf Hafnium 178	73 Ta Tantalum 181	74 W Tungsten 184	75 Re Rhenium 186	76 Os Osmium 190	77 Ir Iridium 192	78 Pt Platinum 195	79 Au Gold 197	80 Hg Mercury 201		81 Tl Thallium 204	82 Pb Lead 207	83 Bi Bismuth 209	84 Po Polonium (210)	85 At Astatine (210)	86 Rn Radon (222)
7	87 Fr Francium (223)	88 Ra Radium (226)	89 Ac▲▲ Actinium (227)	104 Rf Rutherfordium (261)	105 Db Dubnium (262)	106 Sg Seaborgium (263)	107 Bh Bohrium (264)	108 Hs Hassium (269)	109 Mt Meitnerium (268)	110 Uun Ununnilium (269)	111 Uuu Unununium (272)	112 Uub Ununbium (277)							

▲ **Lanthanide elements**

58 Ce Cerium 140	59 Pr Praseodymium 141	60 Nd Neodymium 144	61 Pm Promethium (147)	62 Sm Samarium 150	63 Eu Europium 152	64 Gd Gadolinium 157	65 Tb Terbium 159	66 Dy Dysprosium 163	67 Ho Holmium 165	68 Er Erbium 167	69 Tm Thulium 169	70 Yb Ytterbium 173	71 Lu Lutetium 175

▲▲ **Actinide elements**

90 Th Thorium 232	91 Pa Protactinium (231)	92 U Uranium 238	93 Np Neptunium (237)	94 Pu Plutonium (242)	95 Am Americium (243)	96 Cm Curium (247)	97 Bk Berkelium (245)	98 Cf Californium (251)	99 Es Einsteinium (254)	100 Fm Fermium (253)	101 Md Mendelevium (256)	102 No Nobelium (254)	103 Lr Lawrencium (257)